Advances in Pollen-Spore Research Vol. 28

ECOLOGY OF POLLEN AND FUNGAL SPORE

Edited by

Dr. A.J. Solomon Raju
Head of the Department
Department of Environmental Sciences
Andhra University
VISAKHAPATNAM 530 003

Today & Tomorrow's Printers and Publishers
4436/7, Ansari Road, Daryaganj, New Delhi - 110 002 (India)
Ph : 23242621; 23241021; Fax 23242621; E-mail: ttpp@vsnl.net

ISBN 81-7019-455-5 (India)
ISBN 1-55528-312-8 (USA)

Published by
Today and Tomorrow's Printers and Publishers
4435-36, 7, Ansari Road, Daryaganj, New Delhi - 110 002 (India)
Ph : 23242621; 23241021; Fax 23242621; E-mail: ttpp@vsnl.net

ANNOUNCEMENT

Authors are requested to contribute original and review research papers pertaining to the subject of the Journal of Advances in Pollen-Spore Research by the end of March 31, 2011 for publication in the 2011 issue which is expected to be published in July 2011. A free copy of the Journal will be sent to the first author of each paper published in the issue. Additional copies of the journal may be ordered with Today and Tomorrows Printers and Publishers, New Delhi.

—Prof. Aluri Jacob Solomon Raju
Chief Editor

CONTENTS

Advances in Pollen Spore Research Vol. XXVIII (2010) : 1-16
Editor : Dr. A. J. Solomon Raju
Today & Tomorrow's Printers and Publishers, New Delhi - 110 002 (India)

BEE FAUNAL DIVERSITY IN THE NILGIRIS PART OF THE WESTERN GHATS, TAMIL NADU

K.R. Sasidharan and C. Kunhikannan

Biodiversity Division

Institute of Forest Genetics and Tree Breeding

Coimbatore-641 002

ABSTRACT

The role of bees as pollinators and their importance in ecosystem functioning is increasingly realised now. But they are encountering various threats in their habitats. At this juncture, it is prudent to understand their diversity, seasonality and floral associations, so as to take up conservation measures. The Nilgiri district of Tamil Nadu is an ecologically important hilly tract of the Western Ghats. The studies conducted on the bee fauna of the district have revealed the occurrence of about 92 species of bees. They fall under 15 genera viz. *Apis, Amegilla, Braunsapis, Ceratina, Chelostoma, Halictus, Heriades, Lasioglossum, Megachile, Nomada, Nomia, Sphecodes, Thyreus, Trigona* and *Xylocopa.* The highest diversity of bees was noticed in the Dry deciduous forests, followed by the Semi-evergreen forests. The Sub-tropical broad leaved hill forests, the Moist deciduous forests and the Swamp forests showed intermediate diversity, while the Montane wet temperate forests (Shola forests) and the Thorn forests exhibited lower diversity. The diversity of bee species in monoculture plantations was found to be low, compared to the natural forests. Studies on the floral association have shown that about 113 species of plants belonging to 93 genera and 42 families are associated with bees in the Nilgiris. The Asteraceae members were found to be highly preferred by the bees, followed by that of Lamiaceae and Acanthaceae. Various threat factors encountered by the bees in the Nilgiris are highlighted and conservation strategies suggested.

Key words: Bees, Nilgiris, Forests.

INTRODUCTION

The role of bees as pollinators and their importance in ecosystem functioning is increasingly realised now. They do immense service to nature by facilitating various plant species to continue their reproductive cycles, so as to ensure survival of plants in the ecosystem. Since most of the tropical forest species are cross-pollinated, they require the pollination services from insects, particularly the bees. Pollination is recognized as an important ecosystem function that is directly linked to biodiversity and global food security and in this context, the contribution of bees towards food production and preservation of plant diversity is significant. The bees also provide honey, which has high nutritious as well as medicinal value. The livelihood of many forest dwelling communities is also closely linked to honey collection from the forest areas. The total number of bee species recorded in the world is about 16, 425 coming under 425 genera (Michener 2000). The Indian bee fauna consists of 748 species, coming under 61 genera, which constitutes about 4.55 per cent of the total number of bee species reported in the World.

Bees are under pressure everywhere, both from the direct impact of pesticides in the environment as well as habitat alteration and destruction. There is recognition of a looming pollination crisis and efforts have been mobilized at several levels to address pollinator management and conservation. On a global level, the United Nations Conservation on Biological Diversity (CBD) has established an initiative (the International Pollinator Initiative or IPI) for the conservation and sustainable use of pollinators.

MATERIALS AND METHODS

The Study Area

The Nilgiri district of Tamil Nadu lies between 11^0 02 N and 11^0 452 N latitudes and 76^0 02 E and $77^0$152 E longitudes. Because of the distinct phytogeographical features, the Nilgiris part of the Western Ghats has been treated separately from the rest of the areas of the Ghats. It is also the meeting place of the Western Ghats and the Eastern Ghats. The elevation of the area ranges from 300 m to over 2600 m above msl. The rain fall also ranges from 750 mm to over 7000 mm. The physiography of the Nilgiris is such as to produce marked difference in the rainfall, temperature, aspect and degree of exposure to winds and soil moisture over quite small areas, resulting in the occurrence of local climates and local temperature effects. This great range in the factors of the habitat produces corresponding diversities in the plant communities that inhabit the tract and in the species of which these communities are composed (Ranganathan 1941). According to Ranganathan (1941), most of the familiar types of South Indian forests are represented in the Nilgiris, together with a few that are peculiar to this hilly tract alone.

The forests of the Nilgiri district fall under three territorial forest divisions, one Tiger Reserve and a National Park and the total extent of forest area of the district is 1304.43 Sq. km. As per the revised classification of Indian forests by Champion and Seth (1968) 9 forest types can be recognized in the Nilgiris. They are: Tropical wet evergreen forests, Tropical semi-evergreen forests, Tropical moist deciduous forests, Swamp forests, Tropical dry deciduous forests, Tropical thorn forests, Subtropical broadleaved hill forests, Montane wet temperate forests and Grasslands.

Survey and sampling

A stratified random sampling with proportional allocation was followed for the survey. The area under different forest types occurring in the district was divided in to 5 Sq. km grids and from these grids, 10 percent was selected for sampling (Table 1). In each grid or sampling location, sampling was done along a transect (Southwood 1966), by laying out rectangular plots of size 50 m x 5 m. The total number of such plots laid out in each sampling location was 50. Bees present in those plots were sampled, during different seasons, using insect collection nets of 25 cm diameter. The insects were anaesthetized using Ethyl Acetate, pinned and preserved and subsequently identified by referring to taxonomists. The frequency of occurrence of different species of bees in each plot was recorded. The bee visitations to flowers of different plant species were tracked and the specimens of these host plants were collected and identified, by referring to floras and the Herbaria.

Table 1. Details of sampling locations selected in the Nilgiri District

Forest type/ plantation	No. of sampling locations selected	Location	Code
Tropical wet ervergreen forests	1	Kil Nadugani	WEG-1
Tropical semi-evergreen forests	5	Ooty main road	SEG-1
		Kolithurai Junction	SEG-2

		Field No.12, Kannampilly	SEG-3
		Kottamala RF	SEG-4
		Daivamala	SEG-5
Tropical moist deciduous forests	3	Ooty main road, near Burliyar village	MD-1
		Kilkuppu	MD-2
		Compartment-1, Kargudi	MD-3
Swamp forests	1	Ambalakkad	SW-1
Tropical dry deciduous forests	1	Adikombai road	DD-1
Tropical thorn forests	3	Pilasona	TH-1
		Aralatti Kottai	TH-2
		Vazhathottam to Sirur road	TH-3
Subtropical broadleaved hill forests	2	Laws falls	SBH-1
		Katteri farm	SBH-2
Montane wet temperate forests	11	Tiger hill road	MWT-1
		Doddabetta	MWT-2
		Ellakkadu	MWT-3
		Kothanar parai	MWT-4
		Longwood Shola	MWT-5
		Prime Shola & Porthimund Sholas	MWT-6
		Governor Shola	MWT-7
		Thia Shola RF	MWT-8
		Kundah RF	MWT-9
		Naduvattom Shola RF	MWT-10
		Gundada	MWT-11
Grasslands	1	Western catchment	GL-1
Plantations	4	Muthinad (*Eucalyptus globulus*) Kortha and Muthinad (*Acacia mearnsii*) Kargudi (Teak) Kunjappanai (Tea)	PL-1PL-2PL-3PL-4

Analysis

Shannon Index

The species diversity was worked out using Shannon – Weiner formula: $(H') = -\sum_{i=1}^{k} pi \ln(pi)$, where, k is the number of categories and pi is the proportion of observations found in category i.

Simpson Index

The Simpson Index was worked out using the formula: **D = “ (n / N)** ^{2}**, where** n = the total number of organisms of a particular species and N = the total number of organisms of all species. To easily understand the trend in dominance, the reciprocal form of the index, i.e. (1/D) is adopted. This ensures that the value of the index increases with increase in diversity (Magurran 1988).

Species richness index

The number of species at a site, in a region or in a collection is called species richness, which is the simplest and most useful measure of diversity. Margalef's diversity index $(D_{mg)}$ (Clifford and Stephenson, 1975) was worked out using the following formula: $(D_{mg)} = (S-1)/ \ln N$, where 'S' is the number of species recorded and 'N' is the total number of individuals summed over all species.

RESULTS AND DISCUSSION

The total number of species recorded from the study area is 92. Among these 25 insects have been identified up to species level, 59 up to genus level and 8 are considered to be new species (Table 2).

The bees collected belong to 3 families namely, Apidae, Halictidae and Megachilidae. They also come under 15 genera *viz*. Apis (3 species), Amegilla (10 species), *Braunsapis* (5 species), *Ceratina* (5 species) *Halictus* (22 species) *Lasioglossum* (4 species), *Nomada* (1 species), *Nomia* (16 species), *Sphecodes* (4 species), *Thyreus* (2 species), *Trigona* (1 species), *Xylocopa* (11 species), *Chelostoma* (1 species) *Heriades* (1 species) and *Megachile* (6 species). The genus, Halictus contains the maximum number of species.

Table 2. Checklist of bees recorded from the Nilgiris

Family/ Species	Location Code	Date of collection	Knownspecies	Probablenew species	Identifiedup to genuslevel
(1)	(2)	(3)	(4)	(5)	(6)
Apidae					
Apis cerana indica	All locations	All seasons	√		
Apis dorsata	All locations	-do-	√		
Apis florea	All locations	-do-	√		
Amegilla sp. 1	MWT- 11	31.08.06			√
Amegilla sp. 2	WEG-1	12.10.07			√
Amegilla sp. 3	TH-2	31.11.07			√
Amegilla sp. 4	TH-2	31.11.07			√
Amegilla sp. 5	MWT-8	26.02.08			√
Amegilla sp. 6	MD-1	15.03.08			√
Amegilla sp. 7	SW-1	04.09.08			√
Amegilla sp. 8	SEG-2	18.09.08			√
Amegilla sp. 9	MD-1	26.12.08			√
Amegilla zonata	MD-1	26.12.08	√		
Braunsapis puangensis	WEG-1	12.10.07	√		
Braunsapis mixta	SEG-2	27.07.06	√		
Braunsapis sp. 1	WEG-1 & MD-2	06.09.08 & 18.09.08			√
Braunsapis sp. 2	SEG-4	11.10.07			√
Braunsapis sp. 3	MWT-4	18.04.08			√
Halictidae					
Ceratina binghami	DD-1,TH-1 & SEG-4	19.6.08, 20.06.08 & 05.09.08	√		
Ceratina heiroglyphica	MWT-4, PL-4, WEG-1, MWT-8, SEG-2, SW-1 & SBH-1	1.9.06, 3.9.07, 9.10.07, 12.10.07, 26.02.08 19.04.08 & 04.09.08	√		
Ceratina propingua	MD-2	19.04.08	√		
Ceratina smaragdula	SEG-4	11.10.07	√		

Ceratina unimaculata	SW-1	09.10.07	√		
Halictus constrictus	MWT-3	17.05.06	√		
Halictus gutturosus	DD-1	20.09.06	√		
Halictus liodomus	MWT-6	08.06.07	√		
Halictus sp. (in between *tristis/ semiaerinus*)	SBH-1	10.06.07		√	
Halictus sp. nr.*catullus*	SBH-2	08.06.08		√	
Halictus sp. 1	PL-4	03.09.07			√
Halictus sp. 2	SEG-4	11.10.07			√
Halictus sp. 3	SEG-4	11.10.07			√
Halictus sp. 4	WEG-1	12.10.07			√
Halictus sp. 5	DD-1	19.06.08			√
Halictus sp. 6	SBH-2	08.06.08			√
Halictus sp. 7	MWT-6	20.2.08			√
Halictus sp. 8	MWT-6	20.2.08			√
Halictus sp. 9	SBH-1	4.12.08			√
Halictus sp.10	SW-1	04.09.08			√
Halictus sp.11	WEG-1	06.09.08			√
Halictus sp.12	WEG-1	06.09.08			√
Halictus sp.13	WEG-1	06.09.08			√
Halictus sp.14	WEG-1	06.09.08			√
Halictus sp.15	SEG-5	25.09.08			√
Halictus sp.16	SBH-1	4.12.06			√
*Halictus sp.*17	SBH-1	4.12.06			√
Lasioglossum sp. 1	MWT-5, PL-4, SEG-3 & SBH-2	25.7.06, 3.9.09, 10.10.07, 8.6.08		√	
Lasioglossum sp. 2	MWT-7 & SBH-2	9.6.07 & 8.6.08			√
Lasioglossum sp. 3	TH-2	31.8.07			√
Lasioglossum sp. 4	SBH-1	4.12.08 & 5.12.08			√
Nomada sp.	DD-1	19.6.08			√
Nomia carinata	SEG-2	27.7.06	√		
Nomia sp., nr. *carinata*	MWT-4	01.09.06		√	
Nomia sp. nr. *clypeata*	DD-1	19.06.08		√	

Nomia formosa	MWT-3	17.5.06	√		
Nomia pilipes	TH-2, SEG-4 & DD-1	3.8.07, 11.10.07 & 28.11.07	√		
Nomia sp. nr. *pilipes*	SEG-1	08.06.08		√	
Nomia sp. 1	MD-3	30.8.07			√
Nomia sp. 2	SW-1	9.10.07			√
Nomia sp. 3	SEG-4	11.10.07			√
Nomia sp. 4	DD-1	28.11.07			√
Nomia sp. 5	DD-1	28.11.07			√
Nomia sp. 6	SBH-1	14.3.08			√
Nomia sp. 7	SW-1	4.9.08			√
Nomia sp. 8	WEG-1	6.9.08			√
Nomia sp. 9	WEG-1	6.9.08			√
Nomia sp. 10	MWT-4	17.9.08			√
Sphecodes sp. nr. *fumipennis*	SEG-4	11.10.07		√	
Sphecodes sp.1	MWT-7	9.6.07			√
Sphecodes sp.2	SBH-2	8.6.08			√
Sphecodes sp.3	PL-2	26.9.08			√
Thyreus histrio	MD-1	15.3.08	√		
Thyreus smithii	WEG-1	12.10.07	√		
Trigona iridipennis	DD-1 & TH-3	1.9.07 & 19.6.08	√		
Xylocopa latipes	SBH-1 & SEG-3	14.3.08 & 10.10.07	√		
Xylocopa pictifrons	WEG-1	12.10.07	√		
Xylocopa sp. nr.*amethystina*	TH-1	20.6.08		√	
Xylocopa sp. 1	MWT-6	20.02.08			√
Xylocopa sp. 2	GL-1	22.2.08			√

Xylocopa sp. 3	WEG-1	6.9.08			√
Xylocopa sp. 4	MWT-6	24.9.08			√
Xylocopa sp. 5	MWT-10	25.9.08			√
Xylocopa sp. 6	MWT-10	25.9.08			√
Xylocopa sp. 7	MWT-6	24.9.08			√
Xylocopa sp. 8	SBH-1	4.12.08			√
Megachilidae					
Chelostoma sp.	PL-4	3.9.07			√
Heriades parvulus	MWT-6	9.6.07	√		
Megachile fraterna	DD-1	28.11.07	√		
Megachile hera Bingham	SEG-2	18.9.08	√		
Megachile sp. nr *chlorigaster*	MD-3	29.11.07		√	
Megachile sp.1	SEG-5	25.9.08			√
Megachile sp. 2	DD-1	28.11.07			√
Megachile sp.3	DD-1	19.06.08			√
Total No. Species	25	8		59	
Grand total			92		

Note: - Due to lack of specialized taxonomists on various groups, many of the specimens could be identified only up to genus level and efforts are continued to identify them up to species level.

Batra (1977) collected 97 species during a year at Ludhiana, Punjab. Linsley (1958) collected 439 species during several years at Riverside, California. Fierros-Lopez (1998) collected 172 species of bees coming under 58 genera and 8 families, from the Volcano of Tequila, Mexico, during March 1994 and February 1995. Alves-dos-Santos (1999) recorded 292 species of bees from the Atlantic rain forest, "restinga" and dunes of the north coast of Rio Grande do Sul, Brazil. It was also found that Halictidae was one of the dominant families represented in the above areas. Karunaratne and Edirisinghe (2006) recorded 148 species of bees, coming under 38 genera and 4 families, from 29 collection sites distributed among 13 districts of Sri Lanka. The families, Halictidae and Apidae were found to contain maximum number of species in Sri Lanka.

The Shannon Weiner index, Simpson index and the species richness index (i.e. Margalef's index) of bees worked out for different locations are given in Table 3. The dry deciduous forests were found to have the maximum diversity (Shannon index: 2.0389; Simpson index: 4.8338 and Margalef's index: 3.2415). Some of the locations of Montane Wet Temperate forests (MWT) and the plantations of *Eucalyptus globulus* (PL-1) and *Acacia mearnsii* (PL-2) had a low diversity (i.e. less than 1) compared to other forest types. Binoy *et al.* (1999) observed a progressive reduction in the number of macrohymenopteran species, including bees, with increasing altitude. They observed that, vegetation and microclimatic conditions characteristic at higher elevations influenced the distribution of insects. The montane wet temperate forests and high altitude grasslands at an altitude of around 2000m contained only less number of species. Among the monoculture plantations, the Tea planation (PL-4), showed a higher diversity value (Shannon Index: 1.7412 and Simpson Index: 4.5213) due to the surrounding refuge vegetation – a preferred habitat for bees. A positive correlation between surrounding refuge environment on species diversity of Apoidea has been documented. Such refuge environments are also important places for survival of bee fauna and with reference to the faunal diversity, they are similar to natural biotope (Banaszak, 1996).

The bee fauna of the Oriental Region (includes Indian Sub-continent) is considered to have the fewest species and genera in the world (Michener, 1974) which has been attributed to the abundance of honey bees in this region. Bees are usually numerous in arid or semi-arid regions (Batra, 1977). Fierros-Lopez (*loc. cit.*) during the study conducted at the Volcano of Tequila, Mexico found that the tropical deciduous forests and mixed deciduous forests had high Shannon indices of 3.93 and 3.86 respectively. Similarly, in the present study, maximum diversity was found in the dry deciduous forests.

Floral relationship of bees

The total number of bee host plants recorded during this study was 113 species, belonging to 93 genera and 42 families (Table 4). The Asteraceae (No. of genera: 16 and No. of species: 17) members were found to be highly attractive to the bees, followed by Lamiaceae (No. of genera: 6 and No. of species: 10) and Acanthaceae (No. of genera: 5 and No. of species: 9). Among the Asteraceae members, *Bidens pilosa* L., var. *minor,* a plant of South American origin, introduced and naturalized in open forest areas of the Nilgiris was highly attractive to bees, which supported 17 species of bees. Karunaratne and Edirisinghe (*loc. cit.*) found that, the family Fabaceae consisted of the maximum number of bee host plants (23 species). They observed that, apart from the native plant species, the naturalized exotic flora also equally served as bee floral hosts. The most attractive bee floral host in Sri Lanka was found to be *Hyptis suaveolens* (Lamiaceae), an exotic (native to Central America) weed, naturalized in many parts of the country, which supported 60 species of bees.

Table 3. Diversity indices of bees in different forest types / locations in the Nilgiris

Sl. No.	Forest / vegetation type	Location code	Shannon Weiner Index	Simpson Index (1/D)	Margalef's index
1	Montane wet temperate forests	MWT-1	0.5623	1.6000	0.4809
		MWT-2	0.2868	1.1803	0.4024
		MWT-3	0.9557	2.3333	1.0278
		MWT-4	1.5008	2.7824	2.0055
		MWT-5	0.1732	1.0868	0.3147
		MWT-6	1.5492	3.8923	1.6590
		MWT-7	0.7304	1.7142	0.7711
		MWT-8	0.9363	1.8901	1.1545
		MWT-9	0.3957	1.2384	0.5498
		MWT-10	0.7587	1.5503	0.9927
		MWT-11	1.5053	3.4948	1.6157
2	Tropical moist deciduous forests	MD-1	1.0664	1.8648	1.5903
		MD-2	1.2258	1.9618	2.1193
		MD-3	1.1907	2.4407	1.5733
3	Tropical dry deciduous forests	DD-1	2.0389	4.8338	3.2415
4	Subtropical broadleaved hill forests	SBH-1	1.2719	2.5901	1.9208
		SBH-2	1.3121	2.3628	1.8706
5	Tropical semi-evergreen forests	SEG-1	1.8350	4.7407	2.3083
		SEG-2	0.9618	1.9221	1.3670
		SEG-3	1.0462	2.1110	1.6485
		SEG-4	1.8610	4.3032	2.9262
		SEG-5	1.4638	2.9912	1.8416
6	Swamp forests	SW-1	1.2454	2.7035	1.5924
7	Tropical thorn forests	TH-1	0.8836	1.8933	1.0983
		TH-2	1.4899	2.9641	1.9460
		TH-3	0.8292	1.7234	1.0028
8	Tropical wet evergreen forests	WEG-1	1.4498	3.2440	1.5346
9	Grasslands	GL-1	1.4660	3.6344	2.1879
10	Plantations	PL-1	0.5235	1.5158	0.2212
		PL-2	0.8124	1.7129	0.5926
		PL-3	0.6931	2.0000	1.4427
		PL-4	1.7412	4.5213	2.2325

Table 4. Number of plant families, plant species and associated bee species recorded in the Nilgiris

S.No.	Plant family	No. of plant genera	No. of plant species	No. of associated bee species
1	Acanthaceae	5	9	12
2	Amaranthaceae	2	2	3
3	Anacardiaceae	1	1	1
4	Asclepiadaceae	1	1	1
5	Asteraceae	16	17	24
6	Balsaminaceae	1	1	1
7	Berberidaceae	1	1	1
8	Boraginaceae	1	1	1
9	Cactaceae	1	1	5
10	Caesalpiniaceae	1	1	6
11	Celastraceae	1	1	2
12	Cucurbitaceae	2	2	5
13	Convolvulaceae	3	5	12
14	Erythroxylaceae	1	1	3
15	Euphorbiaceae	3	3	6
16	Fabaceae	5	5	8
17	Gesneriaceae	1	1	1
18	Hypericaceae	1	1	2
19	Lamiaceae	6	10	12
20	Liliaceae	1	1	1
21	Malvaceae	1	2	6
22	Melastomataceae	3	3	6
23	Meliaceae	1	1	1
24	Mimosaceae	2	2	11
25	Myrtaceae	2	3	2
26	Oleaceae	2	2	4
27	Onagraceae	1	1	5

28	Oxalidaceae	2	2	2
29	Passifloraceae	1	1	1
30	Polygonaceae	2	2	3
31	Rhamnaceae	1	1	2
32	Rosaceae	1	3	3
33	Rubiaceae	8	9	11
34	Rutaceae	2	2	4
35	Sapindaceae	2	2	2
36	Solanaceae	1	3	4
37	Sterculiaceae	1	1	2
38	Theaceae	1	1	2
39	Tiliaceae	1	2	2
40	Vacciniaceae	1	1	1
41	Verbenaceae	2	3	9
42	Zygophyllaceae	1	1	6
	Total	93	113	

CONCLUSIONS

Bees are highly sensitive to habitat destruction/ alterations and climate change. The number of solitary bees is declining, particularly in areas of intensive agriculture through loss of nesting habitat and of floral diversity. Ground nesting species seem to be more threatened than those nesting in wood or plant stems (Westrich 1989; Rasmont *et al.* 1992). Because of the importance of wild bees as pollinators of forest plant species and agricultural/horticultural crops, efforts should be made to manage, increase and conserve them. *Apis cerana indica, Apis dorsata* and *A. florea* colonies may be least disturbed by adopting rational honey collection. It may be noted that, their value as pollinators probably far exceeds the value of honey (Batra 1977).

Species of *Xylocopa, Ceratina, Braunsapis* and *Megachile* may be maintained by providing them with suitable nesting material in or near the fields to be pollinated. They nest in thatched roofs, bamboo or wooden beams, dead trees, weed stalks or bushes and reed fences or stakes. Small, undisturbed weedy or bushy areas, where native wild plants grow are helpful. Special hive-like shelters containing materials can be made and have been successfully used on a large scale in Japan, the USA and elsewhere (Bohart 1972).

The bees that nest in earth or ground, such as species of *Nomia, Halictus, Lasioglossum* and *Amegilla* are best maintained by leaving their nesting sites undisturbed. They usually nest in bare sunny areas. Most of the forest areas in the Nilgiris are grazed by the cattle. They graze/ browse many of the herbaceous and shrubby plants, which serve as nectar and pollen sources of bees. The cattle also trample the ground and make it hard. Hence the forests may be protected from cattle grazing, so as to ensure enough food materials and ideal nesting sites to the bees.

The extent of monoculture plantations in the Nilgiri district is 30,697.03 ha, which constitutes about 22.76 per cent of the natural forest areas. Since the diversity of bees are very low in the plantations, the existing monoculture plantations, particularly of exotic species are to be slowly replaced with local indigenous species, so as to create ideal habitats for bees to sustain.

On an average, about 370 ha of the forest areas in the Nilgiris are affected by fire annually. The fire usually destroys the bushy undergrowth - the nesting place of several species of bees. It also cause havoc to the herbaceous and shrubby species, which serve as pollen and nectar sources for the bees. As the fire increases the temperature on the ground, the ground dwelling bee species are also killed. The incidence of fire is likely to be higher due to climate change and hence the fire protection measures are to be given utmost importance.

Pesticides have been one of the most important factors for the decline of bee populations throughout the world. Plowright *et al.* (1978) reported that, aerial spraying of fenitrothion carried out in the forest areas of New Brunswick in 1976 had high impact on the Bumble bees. Fenitrothion sprayed from the air at 0.21 kg/ha caused mortalities ranging from 100 per cent among experimentally caged adults of Bumble bees in exposed habitats down to 47 per cent in cages placed under forest canopy. They also noticed that the pesticide application had a negative impact on the pollen collection by the bees, thereby reducing the chances for cross-pollination and good seed setting. In the Nilgiri district, pesticides to the tune of around 600 metric tonnes in the form of granules and powder and around 28,000 litres in the form of liquids are used per year (Daniels 1996). If at all pesticides are to be necessarily applied, it may be preferably applied as sprays (not as dusts) in the late afternoon or evening or on cool, cloudy windless days, when bees are least active. Pesticides with less residual toxicity have also to be chosen for application, in order to protect the bee population. The usage of chemical pesticides are to be bare minimum, by adopting Integrated Pest Management Strategies and only safer chemicals have to be used, so as to protect and conserve the bees.

The introduced honey bee (*Apis mellifera*) also poses a high threat to native bees. Though the European honey bee (*Apis mellifera*) was not recorded from the Nilgiris, during the present study, their colonization in this area in future cannot be ruled out. Kato *et al.* (1999) while studying the impact of the introduced honey bee, *Apis mellifera*, on native bee communities in the Bonin Island found that there was competition for nectar and pollen of native flowers, between the introduced honey bee and native bees and it favoured the introduced bees on the disturbed islands, which are thoroughly invaded by alien nectariferous, sometimes aggressive weedy plants. They noticed that, the introduced honey bee strongly depended on the alien plants, especially during winter to spring, when native melittophilous flowers were rare. They also found that, from June to November, the introduced honey bee exhaustively utilized native flowers, which had originally been utilized and pollinated by native bees. Hence the European honey bee should not be introduced to the Nilgiris, so as to conserve and protect the native bees.

REFERENCES

Alves-dos-Santos, I. 1999. The bee fauna and melittophilous plants of the Atlantic rain forest, "restinga" and dunes of the north coast of Rio Grande do Sul, Brazil. *Revista Brasileira de Entomologia* 43 (3-4): 191 – 223.

Banaszak, J. 1996. Ecological bases of conservation of wild bees. In: *The Conservation of Bees* (Eds. A .Matheson, S.L. Buchmann, C.O. Toole, P. Westrich and I.H. Williams) Academic Press : London.

Batra, S. W. T. 1977. Bees of India (Apoidea): Their behaviour, management and key to the genera. *Oriental Insects* 11(3): 289 – 324.

Binoy, C. F., George Mathew, V. V., Sudheendrakumar and Narendran, T.C. 1999. Macrohymenopteran fauna of the Silent Valley National Park, Kerala, India. *Bangladesh Journal of Forest Science* 28 (1): 38 – 46.

Bohart, G. E. 1972. The evolution of parasitism among bees. 41st Faculty Honour Lecture. Utah State University, Logan, 30p.

Champion, H.G. and S.K. Seth 1968. *A Revised Survey of the Forest Types of India.* Govt. of India Press, New Delhi.

Clifford, H. T. and W. Stephenson 1975. *An Introduction to Numerical Classification*, Academic Press, London.

Daniels, R J. R. 1996. The Nilgiri Biosphere Reserve: A review of conservation status with recommendations for a holistic approach to management (Working Paper No.16). UNESCO

(South – South Co-operation Programme), Paris, 48p.

Fierros-Lopez, H.E. 1998. Wild bees (Hymenoptera : Apoidea) of *Volcan de Tequila*, Jalisco, Mexico. *Folia Entomologica Mexicana* 102: 21 – 70.

Karunaratne, W. A. I. P. and J. P. Edirisinghe 2006. Current status and future directions in bee taxonomy in Sri Lanka. In: *The Fauna of Sri Lanka, Status of Taxonomy, Research and Conservation* (ed. N.B. Bambaradeniya). The World Conservation Union (IUCN), Asia Regional Species Conservation Programme, Colombo, pp.12-19.

Kato, M., A. Shibata, T. Yasui and H. Nagamasu 1999. Impact of introduced honeybees, *Apis mellifera*, upon native bee communities in the Bonin (Ogasawara) Islands. *Researches on population ecology* 41 (2): 217 – 228.

Linsley, E.G. 1958. The ecology of solitary bees. *Hilgardia* 27 (19): 543 – 599.

Magurran, A.E. 1988. *Ecological diversity and its measurement*. Croom Helm Ltd., London, 179p.

Michener, C.D. 2000. *The Bees of the World*. The Johns Hopkins University Press, Maryland, 913p.

Michener, C.D.1974. *The social behaviours of the bees*. Harvard University Press, USA, 404 p.

Plowright, R.C. B.A. Pendrel and I.A.McLaren 1978. The impact of aerial fenitrothion spraying upon the population biology of bumble bees in southwestern New Brunswick. *Canadian Entomologist* 110 (11): 1145 – 1156.

Ranganathan, C.R. 1941. *Working Plan for the Nilgiris Division*. Govt. Press: Chennai, 202p.

Rasmont, P., C. Gaspar, Q. J. Lecler, A. Jacob-Remade and A. Pauly 1992. The faunistic drift of Apoidea in Belgium. In: *Bees for Pollination* (ed. E. Bruneau). Proceedings of the workshop in the community programme of research and technological development in the field of competitiveness of agriculture and management of agricultural resources (1989-1993), Brussels, 1-2 March, 1992. CEC, Luxembourg, pp. 67-87.

Southwood, T.R.E.1966. *Ecological Methods*. Methuen and Company Ltd: London, 391p.

Westrich, P. 1989. *Die Wildbienen Baden – Wurttembergs*. Ulmer, Stuttgart.

Advances in Pollen Spore Research Vol. XXVIII (2010) : 17-25
Editor : Dr. A. J. Solomon Raju
Today & Tomorrow's Printers and Publishers, New Delhi - 110 002 (India)

SIZE OF POLLEN FORAGING POPULATION AND ITS DYNAMICS IN *APIS CERANA INDICA* AND *APIS MELLIFERA* IN DIFFERENT REGIONS IN KARNATAKA

A. Nagarathna

S. R. N. Adarsh College,

Bangalore 560 018

and

M.S. Reddy

Centre for Apiculture Studies, Department of Zoology,

Bangalore University, Jnana Bharathi, Bangalore 560 056

ABSTRACT

Studies on the foraging population and its dynamics in *Apis cerana indica* and *Apis mellifera* has been carried out in different eco-habitats of Karnataka during 2006-2008. Information pertaining to pollen foragers during different hours of the day and its population dynamics has been obtained through periodical field observations from different apiaries of both *A.cerana* and *A. mellifera*. It was found out that the peak pollen foraging activity in *A.cerana* was noticed during 0700-1100 h with a smaller peak between 1400-1600 h of the day, however in case of *A.mellifera* the onset of foraging behaviour starts a bit late in the day and the peak foraging behaviour was noticed between 0830-1130 h during pre noon and with a smaller peak during 1430-1630 h during afternoon. The data revealed that there is tremendous scope, application and commercially exploit different types of pollen. This has great implication in commercially exploiting pollen as the bee product. The study also revealed that in addition to the honey produced by the bee colony it has great potential and economically viable beekeeping industry pollen as such has nutritive product which is being applied as an alternative dietary food product.

Keywords: Pollen, foraging behaviour, honeybee.

INTRODUCTION

Honeybees are used as agents in cross pollination for increasing production of seeds and fruits in crops, which has been gaining momentum during the past few decades. Honeybees exhibit different foraging behaviour depending upon the floral structure and location of floral rewards (Punchihewa *et al.* 1985; Raju 1990; 2001). Honeybees have proved to be predominant pollinating agents of oil seeds, pulses, fruits and vegetable crops (Singh 1997; Ahmed and Rehman 2002; Kumar *et al.* 2005). The studies were also conducted by Abrol *et al.* (2002) at Jammu to determine the pollinating insects visiting flowers of temperate fruit plants during Februar-March 2002 which revealed that honeybees, *Apis dorsata F., Apis mellifera L. Apis cerana F.* and *Apis florea F.* comprised more than 80 percent of the total flower visiting insects. Honeybees are the potential pollinators of the crop because of their specific instinctive behavioral features for pollen and nectar collection and efficiency in pollen transfer (Crane 1990; Free 1993; Priti and Sihag 1997). Chaudhary *et al.* (2001) showed the increase in fruit set in Litchi was 252.85 percent in open pollination and 336.72 percent more in bee pollinated crop plants.

Honeybees are social insects and are distributed throughout the world with which man has established a harmonious co-existence (Mishra 1995; Nagaraja and Reddy 1996). These insects are of great importance because they not only produce honey, but also are also important pollinating agents for many plants (Kapil and Kumar 1975; Jadhav 1981; Rao *et al.* 1981; Swaminathan and Bharadwaj 1982; Abrol and Bhat 1985; Deshmukh *et al.* 1985; Winston 1987; Rao and Suryanarayana 1988; Woo *et al.* 2000).

Foraging is a social enterprise in which bees collect pollen, nectar, water and propolis from plants (Seeley 1985). The act of collecting all these is called foraging and the bee is a forager. The foraging behaviour of the honeybee is dependent on its innate capacity and responses to the various environmental stimuli. The foraging behaviour is an important aspect of their biology that enables them to adapt themselves to the available vegetational and climatic conditions. When the bee begins to forage, it will usually make several trips per day when the weather conditions are favourable. A single foraging trip may last up to an hour or more, with the pollen-gathering trip often making more time than a nectar-gathering trip. Bees utilize pollen and nectar as food source and nectar is converted into honey and pollen is used to build combs.

The foraging efficiency of honeybee largely depends on the availability of bee forage, conditions of the colony and the foraging range of worker bees. Studies on the foraging activity of bees in different seasons of the year give an indication of the adaptability of the bees in exploiting the bee forage in a locality. Similarly, the extensive knowledge of the pollen sources helps the beekeepers to exploit the sources to the maximum extent so as to develop stronger colonies that are highly desirable from the point of their productivity.

In India, beekeeping industry has evolved from various practices. Some of these practices include bee pollination, honey production and wax production. Use of proper size hives, determined on the basis of the egg laying capacity of the queen and utilization of space for brood rearing and storing of honey, can be ensured with careful management and the survival of the bees throughout winter and summer, in both the plains and hills. Recently beekeeping has become a fascinating and a profitable business (Gatoria *et al.* 2004). Extensive investigations have been conducted on the physiology and learning behaviour of bees (Bertholf 1931; Hertz 1933; Bermant and Gary 1966; Wehner 1967; Wells and Wehner 1971; Caron and Morse 1972).

Flowers are the main stay of bees' life. From flowers they obtain pollen, the protein rich food used mostly to feed the brood, and nectar the carbohydrate fuel for flight, foraging, hive activity and for developing the brood. The bees are automated micromanipulators. These micromanipulators also work to bring about pollination that is absolutely required for the economically important plants.

Many agricultural crop plants are pollinated by honeybees and it is estimated that one third of the total human diet is supplied by cultivated plants that are dependent full yield and quality depends on the activities of bees (McGregor 1976). Several workers have reported studies on the foraging behavior and pollination activities (Verma and Dulta 1986; Rana 1989; Rao 1991; Jhajj *et al.* 1996; Rana *et al* 2003; Belvad 2005).

MATERIALS AND METHODS

The foraging activity of *A. cerana* and *A. mellifera* was observed during different hours of the day and during different days in a month and throughout the year during 2006-2008. The number of bees foraging for pollen was recorded

(Zaitoun and Vorwohl 2003). The foragers were evaluated for pollen and nectar load. The returning foragers were captured at the hive entrance to collect the pollen loads. The foraging activities of *A. cerana* were observed during different hours of the day at regular intervals and throughout the year at different apiary sites. To determine the pollen load per trip, individual foragers were marked on the dorsal side of their thorax and the returning marked pollen foragers were trapped. Number of bees foraging for pollen was recorded at different apiary sites. Apiary sites identified in Karnataka for the study are provided in Figure 1.

Table 1. Different Apiary sites identified in Karnataka for the study

Location	Apiary sites	Code
1	Heserghatta Village	HG
2	Jnanabharathi	JB
3	Arkavathy madhuvana	AK
4	Honnavar	HN
5	Bhagamandala	BG
6	Kuntanahalli	KT
7	Shivakote	SK
8	Shivanahalli	SH

Foraging efficiency of worker bees was measured by counting the number of pollen foragers over 5 minute periods at the entrance of the hive during different hours of the day from initiation to the cessation of foraging activity.

RESULTS AND DISCUSSION

At the apiary site L1 HG shows the number of foragers of *A. cerana* which was very less in the morning gradually increased during the later part of the day and the peak was observed between 1200 to 1300 h which reduced in the mid afternoon to reach a second peak between 1600 to 1700 h and decrease thereafter. But in case of *A. mellifera,* the number of foragers was very less in the morning, however and gradually increased by noon. Further, it was almost same in the mid afternoon and only decreased after 1400 to 1500 h. At the apiary site L_2

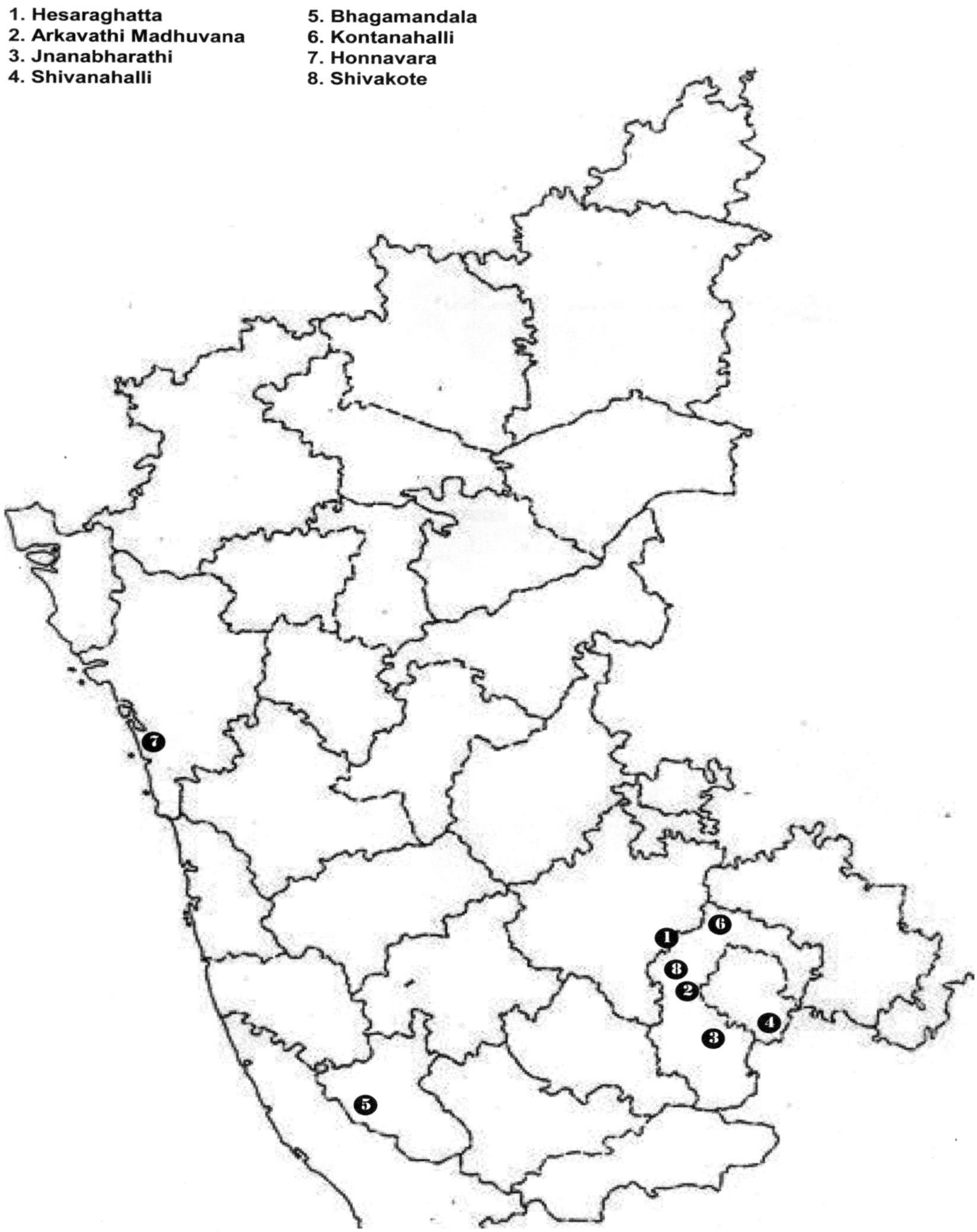

Figure 1. Map of Karnataka showing different Apiary sites

JB shows the size of foraging population was very less in the morning between 0600 to 0700 h and gradually increased in both the species, however, a peak number of foragers were recorded between 1600 to 1700 h for both the species. At apiary site L3AK shows that foraging population was less in the morning and increased gradually in both *A cerana* and *A. mellifera* around mid noon and between 1400 to 1500 h in *A. cerana* and further increased between 1600 to 1700 h in both the species finally both decrease by late evening. The size of foraging population at apiary site L4HN shows the activity of *A. cerana* and *A. mellifera* almost commenced at the same time at 0600 to 0700 h only and continued its foraging activity up to the mid noon and thereafter decreases to cease at 1700 to 1800 h. At the apiary site L5BG the activity of the foraging in both the species *A. cerana* and *A. mellifera* was almost the same with a peak around 1200 to 1300 h, however in case of *A. cerana* there was a second peak was observed during 1600 to 1700 h and thereafter the bees ceased foraging. Size of foraging population of apiary site L6KT showed a steady increase in the morning however, the peak pollen collection was observed only at 1600 to 1700 h otherwise the number of bees carrying pollen or nectar in both *A. cerana* and *A. mellifera* was moderate. L7SK showed the size of foraging population at the location where the size of foragers varied with least early in the day with a gradual increase during the mid noon and there after if continued to rise before it actually ceases to function. This was reported in both *A. cerana* and *A. mellifera*. The study of the size of foraging population however concludes that the number of foragers for pollen or nectar were more for *A. cerana* than *A. mellifera* though both the species were active throughout the day at different apiary sites during the entire study period during 2006-08. Bees collected both pollen and the nectar from the flowers. The foraging rate was high at mid day as compared to the morning and evening. The maximum visitation frequency of insect for pollination was during the mid flowering period, while it was lowest during the beginning and end of the flowering period. As usually the most flowers bloom during late morning and close by evening. The population of foragers collecting pollen was not constant throughout the day. The result concludes that, though both species were active throughout the day, the size of foraging population in *A. cerana* was more than *A. mellifera.* Higher proportion of pollen foraging by the honeybee species was explained on the basis of flowering in the particular area (Thakur *et al.* 1995).

ACKNOWLEDGEMENT

The authors are thankful to Dr. M. Ramachandra Mohan, Professor and Chairman, Department of Zoology, Bangalore University, Jnana Bharathi,

Bangalore for providing the necessary research facilities and the encouragement to carryout this work as Research scholar.

REFERENCES

Abrol, D. P. and Bhat, A.A. 1985. Analysis of biophysical interaction in causing foraging behaviour of some bees. *Indian Bee J.* 47: 36-44

Abrol, D. P., Khan, M.S. and Monobrullah, M. 2002. Competition of fruit trees with a weed, Lepidagthus incurve for their pollination by *Apis species* in Jammu. *Indian Bee J.* 64 (1&2) 55-56.

Ahmed, B and Rehman, A. 2002. Population dynamics of insect foragers and their effect on seed yield of rape seed (*Brassica Compestris l. Van Toria) Indian Bee J.* 64 (3&4): 1-5

Bertholf, L. M.1931. Reactions of honeybees to light. *J.Agri. Res.* 42: 379-419.

Bermant, G. and Gary, N.E. 1966. Discrimination training and reversal in groups of honeybees. *J. Apic. Res.* 5: 179-180.

Belvadi. V., Parvathi, C, Venkateshmurthy, P and Honnalingappa. 2005. Significance of flower visitors for pollination and fruit set in coffee. *Indian Bee J.* 67 (3&4):113-117

Chaudhary, D. K., Singh, B., Singh, A. K. and Singh, P. P. 2001. Influence of honeybee pollination on fruit set and yield of litchi. *Indian Bee J.* 63(1& 2) 54-56.

Caron, D. N. and Morse, R.A. 1972. Attraction of Japanese beetles traps to honeybees, bumblebees and other apoidea. *Envt. Ento.* 1 (3): 272-274

Crane Eva. 1990. Bees and Beekeeping. Bath Press Ltd. Avon 68-69

Deshmukh, A. K., Rao, G. M. and Karve, A. D. 1985. Studies on the honeybee pollination and the yield of sunflower. *Indian Bee J.* 47: 1-4.

Free J. B. 1993. Insect Pollination of Crops. *3rd Edn. Academic Press,* London.

Gatoria, G. S., Pradeep K. Chhuneja, Jaspal Singh and Baljit Singh. 2004. Beekeeping in India: 1. Historical perspective and prospects with special reference to *Apis melliferaL. Indian Bee J.* 66 (3-4): 125-135.

Jhajj, B. S., Gatoria, G. S., and Bakhetia, D.R.C. 1996. Three decades of beekeeping research in Punjab.*Nat. Agri.Tech. Inf centre*,Ludhiana 23-27.

Jadhav, L. D. 1981. Role of insects in the pollination of onion in Maharastra. *Indian Bee J.* 43: 61-64.

Kapil, R. P. and Kumar, S .1975. Foraging activity of *Apis dorsata* on *Brassica juncea.* .Bombay. *J. Natu. Histo Socio.* 7: 327-331.

Kumar Asim Maiti, Pradeep Kumar Singh and Jogeshwari Singh 2005. Floral calendar for beekeeping at the Bichpuri, Agra. *Indian Bee J.* 67(3 and 4): 156-160.

Mc Gregor, S. E. 1976. Insect pollination of cultivated crop plants. *Agri. Hand book U.S Dept.*

of Agri. No.496, Washington, D.C.

Mishra, R. C. and Yogesh Kumar 1995. Introduction and spread of *Apis mellifera* in India and the possibility of inbreeding harm. *Indian Bee J.* 57 (3): 130-134.

Nagaraja N. and Reddy C. C. 1996. Hive and foraging index of hive honeybee species. Spl. issue Proc. of the National Symp. on Ani.Beh. 11: 7-16

Punchihewa, R. W. K., Koeniger, N., Kevan, P.G and Gadawski, R. M. 1985. Observations on the dance communication and natural foraging ranges of *Apis cerana, Apis florea and Apis dorsata* in Sri Lanka. *J. Apic. Res.* 24 (3): 168-175

Priti and Sihag, R. C. 1997. Diversity visitation frequency,foraging mode and pollinating efficiency of the insect pollinators visiting cauliflower *(Brassica olaracea L.* var. Botrytis) blossoms. *Indian Bee J.* 59 (1-4): 1-8.

Raju, A. J. S. 1990 Studies on pollination ecology in India. *A Review. Proc. Indian Nat. Sci. Acad.* 56: 375-388.

Raju, A. J. S and Purnachandra Rao, S. 2001. Forage plants of *Apis cerana indica F.* at Visakhapatnam. India. *Indian Bee J.* 63 (3&4): 21-30.

*Rana B. S.1989. Aggressive and hoarding behaviour of *Apis cerana* and *Apis mellifera* and their role in pollination on plum and apple bloom.Ph.D. Thesis, Himachal Pradesh University.

Rao, G.M. 1991. Studies on the floral biology and pollination requirements of scented methi (*Trigonella corniculata* L.) *Indian Bee J.* 53:39-43.

Rana, B. S., Sud, A. and Rajeshkumar 2003. Role of *Apis mellifera L.* on the pollination of almond *Prunus amygdalus* Batsch in Kotkai, Shimla. *Indian Bee J.* 65 (1&2): 1-5.

Rao G.M, Lazar, M. and Suryanarayana, M. C.1981. Foraging behaviour of honeybees on Seasum. *Indian Bee J.* 45: (3&4) : 97-100.

Rao, G. M. and Suranarayana, M. C. 1988. Studies on pollination of watermelon. *Indian Bee J.* 50 (1): 5-8.

Seeley, T. D. 1985. Honeybee Ecology. *Princeton University Press, Princeton,* 201.

Swaminathan, R. and Bharadwaj, S. C. 1982. Bee pollination on sunflower and foraging behaviour. *Indian Bee J.* 44: 33-34.

Singh, Y. R. 1997. Role of honeybee in farm production, agricultural growth and rural reconstruction in India. *Indian Bee J.* 59 (1&2): 24-30.

Thakur, S. S., Dhaliwal, H. S. and Jitendra Kumar 1995. Pollen foraging ecology of *Apis cerana F.* and *Apis mellifera L.* under midhill conditions of Himachal Pradesh, India. *Indian Bee J.* 57 (3): 127-128.

Wehner, R. 1967. Pattern recognition in bees. *Nature.* 21: 1244-124.

Wells, H. and Wehner 1971. The influence of food scent on behaviour of foraging honeybees. *Physio. Zool.* 44: 191-209.

Winston, M. L. 198.7 The Harvard university press, Cambridge pp. 296.

Woo, S., Kun Yong, Ho Cha. and Hawang Yong Kim 2000. Abilities of *Apis mellifera* and *Apis cerana* to pollinate strawberry in plastic house. *Asian Bee J.* 2 (1&2) : 31-35.

Zaitoun Shahera Talat and Vorwohl Guenther 2003. Major pollen plant spp. in relation to honeybees' activity in the Jordanian Desert area. *Int. Jour. Agri.* Biol. 5: (4) 411-415.

Advances in Pollen Spore Research Vol. XXVIII (2010) : 27-31
Editor : Dr. A. J. Solomon Raju
Today & Tomorrow's Printers and Publishers, New Delhi - 110 002 (India)

PREFERENTIAL VISITATION BY DIFFERENT SPECIES OF HONEY BEES ON THREE ELITE GENOTYPES OF SESAME

Nithya Chandran and Shashidhar Viraktamath

Department of Entomology,
University of Agricultural Sciences
Dharwad - 580005, Karnataka

ABSTRACT

Apis dorsata, A. cerana and *A. florea* were the major visitors on three elite genotypes of sesame *viz.* DS-1, E-8 and SSD-1. *A. dorsata* and *A. florea* mean preferred to visit all the genotypes equally but *A. cerana* showed more preference towards DS-1 and E-8 genotypes. However, equal preference to all the three genotypes was exhibited when total *Apis* bees visitation was considered (3.35 to 3.59 bees/5flowers/min). Bee visitation by *A. dorsata* and *A. cerana* on different days of flowering varied significantly while it was uniform on all the days in case of *A. florea* and also in total *Apis* bees.

Keywords: Honeybees, elite genotypes, sesame, foraging activity.

INTRODUCTION

Sesame (*Sesamum indicum* L.) is one of the oldest cultivated crops in the world. In India it is grown in the states of Gujarat, West Bengal, Rajasthan, Tamil Nadu, Orissa, Madhya Pradesh, Andhra Pradesh, Maharashtra, Uttar Pradesh, Punjab and Karnataka. Although sesame is a self pollinated crop high level of heterosis has been reported. The flower structure of sesame facilitates cross pollination. Earlier reports show that there is 0.5 to 65 per cent natural crossing.

The out crossing percentage in sesame genotypes vary from 6.1 to 20.1 per cent. The seed yield in sesame varieties could be increased concurrently from 9.52 to 39.65 per cent (Patnaik *et al.* 2004). In this context, present studies were made on preferential behaviour of different species of honey bees on three genotypes of sesame and the results are presented in this paper.

MATERIALS AND METHODS

The present investigations were carried out at the Main Agricultural Research Station of the University of Agricultural Sciences Dharwad during kharif season of 2008. Three genotypes of sesame namely, DS-1, E-8 and SSD-1 grown in three separate plots of 6 x 80 m by following recommended package of practices by the AICRP on sesame, Dharwad Center, was utilized for these studies. From each plot of genotype three different spots of 6x10 m were selected. From each of such selected spot, five flowers were selected randomly and number of different honey bees visiting these flowers per minute was recorded at 8, 10, 14 and 16 hr of the day. Such observations were made at peak flowering stage of the crop for ten days starting from 2nd day after initiation of flowering. These observations were made on every alternate day. Means of these observations were calculated species wise separately and also of the pooled data. Later each mean was subjected to "x+0.5 transformations. The data was then subjected to ANOVA using MS - Excel programme and means were separated by DMRT.

RESULTS AND DISCUSSION

Preference of *Apis dorsata*: The mean *Apis dorsata* visitation on three genotypes *viz.* DS-1, E-8, and SSD-1 was 2.24, 2.63 and 2.46 bees /5 flowers /minute, respectively but the preference to these genotypes was statistically equal (Table 1). However the mean preference to visit the genotypes was significantly more on 4, 8, 10 and 12 days after flowering which was at par with each other (2.33 to 2.89 bees/5 flowers/minute). On the 6th day after flowering less preference was shown to visit the flowers (1.50 bees/5 flowers/minute). The bee visitation differed significantly (1.50 to 3.33 bees/5 flowers/minute) when the interaction between genotypes and days of flowering was considered. The highest bee visitation (3.33 bees/5 flowers/minute) was recorded on SSD-1 variety at 10 days after flowering.

Table 1. Preferential visitation by *Apis dorsata* on three genotypes of sesame

Days after flowering	Number of *A. dorsata*/5 flowers/min in genotype			Mean
	DS-1	E-8	SSD-1	
4 DAF	2.33(1.68)[cd]	2.67(1.78)[abc]	2.00(1.58)[d]	2.33(1.68)[a]
6 DAF	1.50(1.41)[e]	1.50(1.41)[e]	1.50(1.41)[e]	1.50(1.41)[b]
8 DAF	2.42(1.71)[bcd]	3.00(1.87)[ab]	3.13(1.90)[a]	2.85(1.83)[a]
10 DAF	2.17(1.63)[cd]	3.17(1.91)[a]	3.33(1.95)[a]	2.89(1.83)[a]
12DAF	2.78(1.81)[abc]	2.83(1.82)[abc]	2.33(1.68)[cd]	2.65(1.77)[a]
Mean	2.24(1.65)[a]	2.63(1.76)[a]	2.46(1.71)[a]	
	S.Em.±	CD		
Genotypes	0.02	NS		
Days	0.01	0.03		
G×D	0.03	0.09		

Figures in parantheses are "(x+0.5) transformed values: DAF – Days after flowering

Preference of *Apis cerana:* *A. cerana* preferred to visit DS-1 and E-8 significantly more (0.82 to 0.94 bees/5 flowers/minute) followed by SSD-1 (0.53 bees/5 flowers/minute) (Table 2). On different days of flowering the mean preference was significantly higher and at par on 4, 6, 10 and 12 days after flowering (0.66, 1.17, 0.66 and 0.82 bees/5 flowers/minute). However, on 8th day significantly lesser preference by bees to visit the flowers (0.50 bees/5 flowers/minute) was observed. The interaction between genotypes and days of flowering

Table 2. Preferential visitation by *Apis cerana* on three genotypes of sesame

Days after flowering	Number of *A. dorsata*/5 flowers/min in genotype			Mean
	DS-1	E-8	SSD-1	
4 DAF	1.33(1.35)[b]	0.33(0.91)[d]	0.33(0.91)[d]	0.66(1.06)[ab]
6 DAF	2.00(1.58)[a]	1.50(1.41)[b]	0.00(0.71)[e]	1.17(1.23)[a]
8 DAF	0.38(0.93)[cd]	0.38(0.93)[cd]	0.75(1.11)[c]	0.50(0.99)[b]
10 DAF	0.33(0.91)[d]	1.33(1.35)[b]	0.33(0.91)[d]	0.66(1.06)[ab]
12DAF	0.66(1.08)[cd]	0.55(1.02)[cd]	1.25(1.32)[b]	0.82(1.14)[ab]
Mean	0.94(1.17)[a]	0.82(1.13)[ab]	0.53(0.99)[b]	
	S.Em.±	CD		
Genotypes	0.03	0.07		
Days	0.03	0.10		
G×D	0.02	0.07		

Figures in parantheses are "(x+0.5) transformed values: DAF – Days after flowering revealed that the mean bee visitation was significantly different and ranged from 0.00 to 2.00 bees/5 flowers/minute.

Preference of *Apis florea*: *A. florea* preferred all the three genotypes equally with 0.14 to 0.59 bees/5 flowers/minute (Table 3). Similarly, equal preference to visit the flowers was also shown on different days after flowering (0.11 to 0.68 bees/5 flowers/minute). However, the interaction between genotypes and days of flowering showed that the bee visitation was significant and varied from 0.00 to 1.83 bees/5 flowers/minute.

Table 3. Preferential visitation by *Apis florea* on three genotypes of sesame

Days after flowering	Number of *A. dorsata*/5 flowers/min in genotype			Mean
	DS-1	E-8	SSD-1	
4 DAF	0.44(0.95)^{bc}	0.11(0.78)^{bc}	0.33(0.91)^{bc}	0.29(0.88)^{a}
6 DAF	0.33(0.90)^{bc}	0.00(0.71)^{c}	0.00(0.71)^{c}	0.11(0.77)^{a}
8 DAF	0.00(0.71)^{c}	0.17(0.81)^{bc}	0.58(1.03)^{b}	0.25(0.85)^{a}
10 DAF	0.11(0.78)^{bc}	0.22(0.84)^{bc}	0.22(0.84)^{bc}	0.18(0.82)^{a}
12DAF	0.00(0.71)^{c}	0.22(0.84)^{bc}	1.83(1.50)^{a}	0.68(1.02)^{a}
Mean	0.18(0.81)^{a}	0.14(0.80)^{a}	0.59(1.00)^{a}	
	S. Em. ±	CD		
Genotypes	0.04	NS		
Days	0.05	NS		
G×D	0.08	0.24		

Figures in parantheses are "(x+0.5) transformed values: DAF – Days after flowering

Preference of all *Apis* bees (pooled data of different *Apis* species): The total number of all *Apis* bees showed equal preference to all the three genotypes (3.35 to 3.59 bees/5 flowers/minute) (Table 4). However, higher preference by the bees was recorded on 4th, 8th 10th and 12th days after flowering (3.29 to 4.15 bees/5 flowers/minute). Significantly less number of bees was recorded on 6th day after flowering (2.78 bees/5 flowers/minute). The bees visited differently when the interaction between genotypes and days of flowering was considered (2.61 to 5.42 bees/5 flowers/minute).

Present results indicate no special preference to any particular genotype of sesame by the honey bees. Perhaps flowers of all these three genotypes may offer equal rewards to the honey bees in terms of nectar. Contrarily, Sachdeva *et al.* (2003) reported that *A. dorsata* was more abundant on HT-1 cultivar of

Table 4. Preferential visitation by total *Apis* honey bees on three genotypes of sesame (Pooled)

Days after flowering	Number of *A. dorsata*/5 flowers/min in genotype			Mean
	DS-1	E-8	SSD-1	
4 DAF	4.10(2.14)^{bcd}	3.11(1.90)^{def}	2.66(1.78)^{f}	3.29(1.94)^{ab}
6 DAF	3.83(2.08)^{bcde}	3.00(1.87)^{def}	1.50(1.41)^{g}	2.78(1.79)^{b}
8 DAF	2.79(1.81)^{ef}	3.54(2.01)^{cde}	4.46(2.22)^{abc}	3.60(2.02)^{ab}
10 DAF	2.61(1.76)^{f}	4.72(2.28)^{ab}	3.88(2.08)^{bc}	3.73(2.04)^{ab}
12DAF	3.44(1.98)^{cde}	3.60(2.02)^{bc}	5.42(2.42)^{a}	4.15(2.14)^{a}
Mean	3.35(1.96)^{a}	3.59(2.02)^{a}	3.58(1.98)^{a}	
	S.Em.±	CD		
Genotypes	0.03	NS		
Days	0.04	0.13		
G×D	0.08	0.22		

Figures in parantheses are "(x+0.5) transformed values: DAF – Days after flowering

sesame. However, the variation of preference to visit the flowers on different days of flowering could be due to the differences in the number of fresh flowers available on any particular day. Patnaik *et al.* (2004) has also reported more *A. cerana* bee visitation on sesame flowers 41 days after sowing.

ACKNOWLEDGEMENT

The authors are grateful to Dr. H.G. Palakshappa, AICRP on sesame, UAS, Dharwad, for giving permission to conduct the above studies on three different genotypes of sesame. Acknowledgements are also due to Dr. A.S. Vastrad and Dr. R.K. Patil, UAS, Dharwad for their suggestions in conducting the study. This study was partially funded by the Department of Biotechnology, Ministry of Science & Technology, Government of India, through an adhoc project.

REFERENCES

Patnaik, H. P., Mohapatra, L. N. and Das, B. 2004. Effect of bee pollination on the yield of sesame under protected conditions. *Indian Bee J.* 66: 84-91.

Sachdeva, Y., Bhatnagar, P. and Gulati, R. 2003. Relative abundance and foraging behaviour of *Apis* spp. on sesamum (*Sesamum indicum*) flowers. *Ann. Pl. Protec. Sci.* 11: 281-283.

Advances in Pollen Spore Research Vol. XXVIII (2010) : 33-37
Editor : Dr. A. J. Solomon Raju
Today & Tomorrow's Printers and Publishers, New Delhi - 110 002 (India)

FORAGING BEHAVIOUR OF HONEY BEES ON THREE ELITE GENOTYPES OF NIGER

Nithya Chandran and Shashidhar Viraktamath

Department of Entomology, University of Agricultural Sciences
Dharwad 580005, Karnataka

ABSTRACT

Apis dorsata, A. cerana and *A. florea* were the major flower visitors on three elite genotypes of niger namely, DNC-4, RCR-18 and NO-71. Mean *A. dorsata* and *A. cerana* visitation was significantly different with higher number of bees on RCR-18 (3.29 and 2.34 bees) and No. 71 (2.73 and 3.40 bees) followed by DNC-4. *A. florea* visitation did not differ among the genotypes. *A. dorsata* visitation was higher on 11th, 15th and 17th day after flowering and *A. cerana* was more on 11th and 17th day. *A. florea* visitation was more only on 11th day after flowering. Similarly mean total *Apis* bees visitation was more on 11th day.

Keywords: Honeybees, foraging behaviour, elite genotypes, niger.

INTRODUCTION

India is one of the major countries growing niger (*Guizotia abyssinica* Cass) which accounts for more than 50 per cent of area and production of the world. Niger is self incompatible with 100 per cent entomophilous cross-pollination. Earlier studies have revealed that the yield of niger being reduced by 11 to 78 per cent in the absence of honey bees and natural pollinators. (Anonymous 2005). Constant efforts are being made to increase the productivity of niger by developing new high yielding genotypes. It is of great interest to study the preference of the new genotypes by different species of honey bees. This would also give an insight to the breeders to develop better genotypes, the productivity of which can be enhanced by employing honey bees for pollination. In this context, the present studies were made on the foraging behaviour of honey bees on three elite genotypes namely, DNC-4, RCR-18 and No-71 and the results are presented in this paper.

MATERIALS AND METHODS

The present investigations were carried out at the Main Agricultural Research Station of the University of Agricultural Sciences Dharwad which is located at 15°26' North latitude, 75°07' East longitudes and at an altitude of 678 meters above mean sea level. Three elite genotypes of niger *viz.* DNC-4, RCR-18 and No: 71 grown in three separate plots of 6 x 80 m by following standard package of practices by the AICRP on niger, Dharwad center, during kharif season of 2008, were utilized for the present studies. From each plot of genotype, three different spots of 6x10 m were marked and in each of these spots, five flowers were selected randomly and number of different honey bees visiting these flowers per minute was recorded at 8, 10, 14 and 16 hr that coincided with their peak foraging activity. These observations were made at peak flowering stages for ten days on every alternate day. Observations commenced from seven days after initiation of flowering (52 days after sowing). Means of these observations were calculated and subjected to factorial RBD analysis after "(x+0.5) transformation of the data. The means were separated by DMRT at 5 per cent significance level.

RESULTS AND DISCUSSION

Three species of honey bees *viz. Apis dorsata, A. cerana* and *A. florea* were recorded visiting all the three genotypes of sesame. The mean *Apis dorsata* visitation was significantly different with higher number of bees on RCR-18 and No-71 (3.29 and 2.73 bees/5 flowers/minute respectively) (Table 1). Similarly, *A. cerana* visitation was significantly higher on RCR-18 and NO-71 with 2.34 and 3.40 bees /5 flowers /minute which were on par with each other (Table 2). However, *A. florea* had equal preference to different genotypes of niger though the bee visitation varied from 4.33 to 5.97 bees /5 flowers /minute (Table 3). Total number of all *Apis* bees visitation was on par on different genotypes with 9.13 to 11.60 bees/5 flowers/minute (Table 4).

Table 1. Foraging behaviour of *Apis dorsata* in three genotypes of niger

Days after flowering	Number of *A. dorsata*/5 flowers/min in genotypes			Mean
	DNC-4	RCR-18	NO-71	
9 DAF	0.42(0.96)[i]	1.46(1.40)[gh]	3.96(2.11)[bc]	1.94(1.49)[b]
11 DAF	1.25(1.32)[gh]	6.57(2.66)[a]	1.57(1.43)[fgh]	3.13(1.81)[a]
13 DAF	1.87(1.53)[efg]	1.57(1.43)[fgh]	0.84(1.15)[hi]	1.42(1.37)[b]
15 DAF	2.50(1.73)[def]	3.13(1.89)[cd]	2.92(1.83)[cde]	2.85(1.81)[a]
17 DAF	3.13(1.90)[bcd]	3.75(2.06)[bc]	4.38(2.21)[b]	3.75(2.06)[a]
Mean	1.83(1.49)[b]	3.29(1.89)[a]	2.73(1.74)[ab]	
Genotypes	S.Em.±	CD		
Days	0.01	0.03		
G×D	0.02	0.05		
	0.05	0.16		

Table 2. Foraging behaviour of *Apis cerana* in three genotypes of niger

Days after flowering	Number of *A. cerana*/5 flowers/min in genotypes			Mean
	DNC-4	RCR-18	NO-71	
9 DAF	2.07(1.60)[de]	2.09(1.60)[de]	1.25(1.32)[fg]	1.80(1.51)[bc]
11 DAF	2.82(1.81)[cd]	3.13(1.91)[c]	1.67(1.47)[ef]	2.54(1.73)[ab]
13 DAF	0.25(0.87)[h]	1.88(1.54)[ef]	5.00(2.35)[b]	2.38(1.58)[bc]
15 DAF	0.83(1.15)[g]	2.08(1.60)[de]	2.22(1.64)[de]	1.71(1.47)[c]
17 DAF	1.67(1.46)[ef]	2.50(1.73)[cde]	6.88(2.70)[a]	3.68(1.96)[a]
Mean	1.53(1.38)[b]	2.34(1.68)[a]	3.40(1.90)[a]	
Genotypes	S.Em.±	CD		
Days	0.01	0.03		
G×D	0.05	0.14		
	0.09	0.25		

Table 3. Foraging behaviour of *A. florea in* three genotypes of niger

Days after flowering	Number of total *A. florea* 5 flowers/min in genotypes			Mean
	DNC-4	RCR-18	NO-71	
9 DAF	3.75(2.06)[de]	1.67(1.47)[f]	3.05(1.86)[e]	2.82(1.80)[c]
11 DAF	9.38(3.14)[a]	9.17(3.11)[a]	6.56(2.66)[b]	8.37(2.97)[a]
13 DAF	6.57(2.66)[b]	10.03(3.24)[a]	2.75(1.80)[e]	6.45(2.57)[b]
15 DAF	5.42(2.43)[bc]	5.83(2.51)[bc]	4.72(2.28)[cd]	5.32(2.41)[b]
17 DAF	3.75(2.06)[de]	3.13(1.90)[e]	4.58(2.25)[cd]	3.82(2.07)[c]
Mean	5.77(2.47)[a]	5.97(2.44)[a]	4.33(2.17)[a]	
Genotypes	S.Em.±	CD		
Days	0.05	NS		
G×D	0.06	0.17		
	0.10	0.29		

Table 4. Foraging behaviour of all *Apis* pollinators in three genotypes of niger (pooled data of three *Apis* species)

Days after flowering	Number of total *Apis* bees/ 5 flowers/min in genotypes			Mean
	DNC-4	RCR-18	NO-71	
9 DAF	6.24(2.60)[f]	5.21(2.39)[f]	8.26(2.95)[e]	6.57(2.65)[c]
11 DAF	13.45(3.73)[bc]	18.87(4.40)[a]	9.80(3.21)[de]	14.04(3.78)[a]
13 DAF	8.69(3.03)[de]	13.48(3.73)[bc]	8.59(3.01)[de]	10.25(3.26)[b]
15 DAF	8.75(3.04)[de]	11.04(3.39)[cd]	9.86(3.21)[de]	9.88(3.21)[b]
17 DAF	8.54(3.01)[de]	9.38(3.14)[de]	15.83(4.04)[b]	11.25(3.39)[b]
Mean	9.13(3.08)[a]	11.60(3.41)[a]	10.47(3.28)[a]	
Genotypes	S.Em.±	CD		
Days	0.05	0.06		
G×D	0.06	0.19		
	0.11	0.32		

The mean *A. dorsata* visitation was higher on 11th , 15th and 17th days after flowering (2.85 to 3.75 bees/5 flowers/minute) (Table 1). Higher number of *A. cerana* bee visits was recorded on 11th and 17th day after flowering (2.54 and 3.68 bees /5 flowers /minute) followed by 9th and 13th day (1.80 and 2.38 bees/5 flowers/minute) (Table 2). *A. florea* visited in higher number on 11th day after flowering (8.37 bees/5 flowers/minute) followed by 13th and 15th day (6.45 to 5.32 bees/5 flowers/minute) (Table 3). When all *Apis* bees were considered, higher mean bee visitation was recorded on 11th day after flowering (14.04 bees/ 5 flowers/minute) followed by 13th , 15th and 17th day with 10.25, 9.88 and 11.25 bees/5 flowers/minute respectively which were *on par* with each other (Table 4). The interaction between the days and the genotypes on mean bee visitation was found to be significant in case of *A. dorsata, A. cerana, A. florea* and total *Apis* bees. These results perhaps indicate that *A. dorsata* and *A. cerana* may get more rewards in terms of nectar and pollen by visiting RCR-18 and No.71 genotypes of niger. Similar results have been reported by Panda *et al.* (1993) who observed that *Apis* bees visitation differed in various cultivars of niger. No definite pattern of bee visitation on different days of flowering was apparent. Since, niger produces abundant flowers over a long period, the bee visitation may depend on the number of fresh flowers available on any particular day.

ACKNOWLEDGEMENT

The authors are grateful to Dr. H.G. Palakshappa, AICRP on sesame, UAS, Dharwad, for giving permission to conduct the above studies on three different genotypes of sesame. Acknowledgements are also due to Dr. A.S. Vastrad and Dr. R.K. Patil, UAS, Dharwad for their suggestions in conducting the study. This study was partially funded by the Department of Biotechnology, Ministry of Science & Technology, Government of India, through an adhoc project.

REFERENCES

Anonymous, 2005. Observations made during “national seminar on strategies for enhancing production and export of sesame and niger” and “Annual Workshop of AICRP on sesame and niger research workers 2005”.

Panda, P., Sontakke, B. K. and Panda, B. 1993. Foraging behaviour of honey bee species on different varieties of niger, *Guizotia abyssinica* Cass. in Orissa, *J. Ins. Sci.* **6**: 104-106.

Advances in Pollen Spore Research Vol. XXVIII (2010) : 39-49
Editor : Dr. A. J. Solomon Raju
Today & Tomorrow's Printers and Publishers, New Delhi - 110 002 (India)

COMPARATIVE ANALYSIS OF AEROSPORA AT FIVE DIFFERENT SITES IN LUCKNOW, INDIA

Asha Khandelwal
3/58 Vikas Nagar, Lucknow

Shilpa Singh, Shantanu Chatterjee
Birbal Sahni Institute of Palaeobotany, Lucknow

Rashmi Tewary
Ram Nivas, Moti Nagar, Lucknow

Lily Misra
Department of Botany, Rajasthan University, Jaipur

Rashmi Saxena
532/497, Chandra Lok Road, Aliganj, Lucknow

Priti Agarwal
86/327, Deo Nagar, Kanpur

and

Alka Srivastava
D.G. Degree College, Kanpur

ABSTRACT

The air-borne pollen grains and fungal spores, the main biocomponents of air are found responsible for the causation of respiratory disorders in sensitive individuals. The air-monitoring in relation to flowering periods, mode of pollination, qualitative and quantitative assessment of bioaerosols are the pre-requisites for allergists to identify the offending aeroallergens followed by correct diagnosis and proper treatment. Here, an attempt has been made to present the comparative analysis of qualitative and quantitative prevalence of air-borne pollen grains trapped from five sites of Lucknow. Burkard personal sampler was used throughout the period of investigation (January, 1997-December 1997). The comparative study exhibited the variation in both qualitative and quantitative incidences of pollen grains at all the sites which appeared to be governed mainly by surrounding vegetation and meteorological conditions. Out of forty four taxa recorded, the pollen grains of families Poaceae and Cheno-Amaranth type contributed largely at all the places, followed by Asteraceae. Due to pollen variation, special emphasis is necessary for periodical air monitoring at different localities of the town in order to provide specific zonal data to local clinicians/ allergists.

Keywords: Lucknow, aerospora, pollen grains, pollen variation.

INTRODUCTION

Air-borne pollen grains, fungal spores and various other bioparticales are known to play important role in respiratory disorders (Lewis 1984; Shivpuri 1980). For effective diagnosis and treatment of these ailments, information on diurnal, seasonal and annual variations in the type and concentration of air-borne pollen grains and fungal spores is a pre-requisite (Smart and Knox 1979). The qualitative and quantitative composition of pollen and fungal spores in the atmosphere varies from place to place depending upon the flora growing in near vicinity, human interference and meteorological factors prevailing at that time. It has been observed that aerobiological surveys are generally localized leading to restricted information. In order to achieve precision in data for extensive area many sites are to be explored simultaneously adopting similar methodology. In the last five decades several pollen and fungal spore calendars of Lucknow have been prepared periodically at Birbal Sahni Institute of Palaeobotany, Lucknow but the air sampling was conducted at single site only. The rapid industrialization and urbanization have changed the phenology, flowering periods and distribution of vegetation. Therefore, to update the knowledge of bioaerosols, the air monitoring was carried out by using Burkard personal sampler for a period of one year (January, 1997-December 1997) at five different ecological sites in Lucknow city. Here, a comparative analysis of both qualitative and quantitative prevalence of air-borne pollen of different sites has been made. In addition, a correlation has been done with qualitative abundance of pollen registered during earlier investigations at Birbal Sahni Institute of Palaeobotany, Lucknow.

MATERIAL AND METHOD

Description of sites:

The district of Lucknow (26°30'-27°10'N and 80°30'-81°13'E) falls in the main gangetic plain of Uttar Pradesh and practically has no natural formations except two patches of forests i.e. Kukrail and Musabagh reserve forests. The family Poaceae is dominant and then followed by Fabaceae, Asteraceae, Cyperaceae, Euphorbiaceae, Acanthaceae, Convolvulaceae, Scrophulariaceae, Amaranthaceae and Lamiaceae (Kapoor 1962) .The following sites have been selected for aerobiological samplings:-

1. **Chikan Work Place, Chowk**:

The whole area is famous for intricate embroidery of chikan work. The area is densely populated where lots of shops, inhabitants and skilled chikan workers are always found working for their livelihood. The old cultural heritage of Lucknow is well preserved by both Muslim ladies and gent's embroidery workers. The preparation of cloth for designing, printing, embroidery and finally the finished articles (Saris, kurta, bed covers, salwar suits etc.) can be seen almost in every corner of the area. The sampling site chosen for aerobiological sampling was printing section, where all the raw materials are brought for printing the designs. The common road side trees of Chowk area are the similar to those trees growing in and around the city such as *Albizzia lebbeck, Azadirachta indica, Cassia fistula, Dalbergia sissoo, Delonix regia, Ficus benghalensis, F. religiosa, Holoptelea integrifolia, Mangifera indica, Pterospermum acerifolium, Eucalyptus* sp., *Syzygium cumini, Tamarindus indica* etc.

2. **Garbage Disposal Unit, Gaughat**:

This place was chosen for aerobiological sampling as all water-borne waste of the city are poured in Gomti river. The trees of *Tamarindus indica, Holoptelea integrifolia, Ficus religiosa, Polyalthia longifolia, Mangifera indica* and *Syzygium cumini* are found in near vicinity.

3. **Military Dairy Farm, Dilkusha**:

This area is known for its farming where crops and vegetables are grown. The sampling was done in nearby field as the entrance in the campus was declared prohibited for the general public. The trees growing in the area are *Acacia arabica, Albizzia lebbeck, Azadirachta indica, Mitragyna parvifolia*. The common shrubby elements which are found dispersed throughout the area are *Adhatoda vasica, Calotropis procera, Datura metal, Jatropha gossypifolia* and *Lantana camara* var. *aculeata*. The common rainy season plants are: *Acalypha ciliata, Cassia occidentalis, Commelina* sp., *Cyperus* spp., *Leucas aspera, Malvastrum coromandelianum, Solanum nigrum, Tephrosia purpurea, Trianthema portulacastrum* etc. and the common winter season herbs such as *Ageratum conyzoides, Anagallis arvensis, Argemone mexicana, Chenopodium album, Eclipta prostrata, Tribulus terrestris* etc. grow abundantly along with grasses.

4. **Vegetable Market, Kaiserbagh**:

This place is known as biggest vegetable and fruit market in the city. The fish and poultry markets are also present across the road. The main source of air-borne fungi are rotten vegetables and fruits, dumps of gunny bags, paper bags, packing materials, straw and discarded leaves and stems. In rainy season the place becomes more vulnerable for the growth and proliferation of aeromycoflora. The stray occurrence of tree has been noticed.

5. **Vivekanand Polyclinic:**

The Vivekanand Polyclinic is situated near I.T. College crossing. It is huge building with different departments for specific diseases. The patients visiting to the polyclinic are generally of middle and lower income group. The lobby of the polyclinic was found to be ideal for indoor sampling where all the patients gather initially for collecting their reference cards. The outdoor sampling was done near the gate. The hospital building is quite neat and clean except spitting corners, bathrooms and a few neglected places which were probably responsible for the development of indoor aeromycoflora. The main road-side trees include *Azadirachta indica, Holoptelea integrifolla, Delonix regia, Ailanthus excelsa, Mangifera indica, Syzygium cumini, Tamarindus indica, Eucalyptus* sp., etc. The common herbs and shrubs of the area are *Lantana camara, Cassia occidentalis, C. tora, Dodonaea viscosa, Croton bonplandianum, Datura metal, Argemone mexicana, Cyperus rotundus*, different kinds of grasses, etc.

Description of air sampler:

In present study, Burkard personal sampler was employed and the slide was exposed weekly for ten minutes at all the sites. It is a compact battery/power operated, 10 cm in height and 8 cm in diameter. It has rectangular orifice at the top and a slit on the side to insert an adhesive coated microslide. The sampler sucks in air at the rate of 10 litres/min. through the orifice and the particles get impacted on the slide in the form of a band. It is used for pollen/fungal identification and estimation. The super deluxe microslide (75 mm x 25 mm) of 0.8 mm thickness is used. The 3/4th of the slide is smeared with safranin stained glycerin jelly in the form of a thin film. The labeled slide is inserted through the slit in the sampler with jelly side up and the lid is twisted in order to seal the inner chamber. After exposure, the slide is removed and brought to the laboratory carefully in the small slide box. For mounting, the jelly is melted on a hot plate and a drop is placed on the exposed band of the slide. Warm 22 sq mm cover glass is placed on the drop of molten jelly. The cover glass is slightly pressed with forceps back to get an even, thin mount and thus spreading impacted material. The entire cover slip area is scanned under the light microscope for identifying and counting the different biota. The counts are expressed as number of spores/pollen per cubic meter of air sampled and conversion formula is used.

The apparatus which was first used by Lakhanpal and Nair (1958) for aeropalynological survey of Lucknow was the one designed and indigenously fabricated by Prof. O.A. Hoeg of Oslo, Norway in 1951 during his tenure as Director of Birbal Sahni Institute of Palaeobotany. This apparatus can be placed in group classified by Gregory (1961) under 'Impaction using wind movement by vertical and inclined microscope slide'. The apparatus has wind-wane and horizontally placed hollow tube in which the slide is kept in slanting position over a stand. The tube protects the slide from rain. The tube is fixed over a stand having a pivotal-end so that the tube rotates according to the wind current. Later on, same air-sampler was used to prepare the pollen and fungal spore calendars of three consecutive years during 1969 and again in 1983. The graphic representation of important pollen grains and fungal spores of year 1969-70, 1970-71 and 1983-86 have been published elsewhere for ready assessment of daily fluctuations in their frequencies (Vishnu Mittre and Khandelwal, 1973; Khandelwal, 1988, 1989, 1992, 2001, 2002). A two-year (1980-81) survey of air-borne pollen alone was carried out, using similar air-sampler in the National Botanical Research Institute, Lucknow, (Chaturvedi *et al.* 1987-88). Thus, the various pollen/spore calendars prepared in Lucknow periodically provided a strong base to local allergist and clinicians for the identification of offending aeroallergens for correct diagnosis and proper treatment.

RESULTS

The aerobiological sampling was done at five different sites in Lucknow, viz. Chikan Work Place, Chowk; Garbage Disposal Unit, Gaughat; Military Dairy

Farm, Dilkusha; Vegetable Market, Kaiserbagh and Vivekanand Polyclinic (Indoor and Outdoor) from January, 1997-December,1997 by employing Burkard personal sampler. The detailed weekly, monthly and annual data of qualitative and quantitative prevalence of pollen grains/fungal spores/fungal colonies from each site has already been reported elsewhere (Anonymous 1994-1998). In the present study, the qualitative and quantitative prevalence of only pollen grains are reported.

Forty-four types of pollen grains were recorded from these five sites. Acanthaceae, Asteraceae, *Azadirachta indica,* Brassicaceae, *Cannabis sativa, Casuarina equisetifolia,* Cheno-Amaranth type, *Datura* sp., *Eucalyptus* sp., Fabaceae, *Holoptelea integrifolia,* Malvaceae, *Melia azedarch, Morus alba,* Poaceae, *Ricinus communis,* Urticaceae and Unidentified pollen grains were reported from all the sites in varying frequencies. *Parthenium hysterophorous* was reported from all the sites except vegetable market. Whereas, *Argemone mexicana,* Apiaceae, *Dodonaea viscose, Heliotropium* sp., *Justicia* sp., *Pinus roxburgii, Syzgium cumini, Terminalia* sp., and *Xanthium strumarium* were registered from three sites and *Ailanthus excelsa, Bougainvillea spectabilis* and Polyads from two sites. There were a few pollen taxa which were recorded from one site only i.e. *Anagallis arvensis, Cassia sp., Solanum* sp., and *Impatiens* sp. from Chikan work place, Chowk; *Aegle marmelos,* Cyperaceae, *Typha* sp., *Zea mays* from Military dairy farm; *Cleome* sp., *Putranjiva roxburghii* and *Jatropha* sp. from Vivekanand Polyclinic and Myrtaceae from Vegetable market. In spring and early summer arboreal pollen grains dominated; in late summer and the rainy season, pollen grains of grasses and weeds were recorded, while in the late rainy season and winter, pollen of grasses and cultivated plants were found. However, there were some pollen grains which occurred throughout the year in small or sporadic numbers. Similar comparative analyses of airspora from two different sites have also been reported from Kerala (Gopi *et al.* 1990), Jabalpur (Oomachan, *et al.* 1988) and Gulbarga (Mari Bhat and Rajasab 1985) wherein the variation in aerospora was assigned to different vegetation pattern. Lacey (1962) carried out a similar kind of investigation at two contrasting adjacent rural sites.

Table 1. Showing the number of pollen grains recorded, using Burkard personal sampler from the different sites of Lucknow.

Pollen grains	SITE 1	SITE 2	SITE 3	SITE 4	SITE 5
Acacia nilotica	-	4	2	-	-
Acanthaceae	3	13	5	7	21
Anagallis arvensis	6	-	-	-	-

Argemone mexicana	18	40	17	-	-
Asteraceae	72	96	109	74	58
Azadirachta indica	6	2	2	13	7
Apiaceae	8	-	-	12	7
Ailanthus excelsa	1	-	-	-	4
Aegle marmelos	-	-	4	-	-
Bougainvillea spectabilis	-	10	-	9	-
Brassicaceae	25	24	34	32	16
Cannabis sativa	27	25	19	11	29
Casuarina equisetifolia	15	26	38	31	52
Cassia sp.	13	-	-	-	-
Cheno-Ams type	243	299	242	317	333
Cleome sp.	-	-	-	-	29
Cyperaceae	-	-	5	-	-
Datura sp.	84	10	17	27	7
Dodonaea viscosa	6	5	7	-	-
Eucalyptus sp.	16	23	25	7	22
Fabaceae	6	89	47	69	20
Heliotropium sp.	5	10	23	-	-
Holoptelea integrifolia	30	46	32	17	24
Jatropha sp.	-	-	-	-	8
Justicia sp.	11	16	12	-	-
Malvaceae	30	18	20	28	13
Melia azedarch	11	24	30	4	14
Myrtaceae	-	-	-	9	-
Morus alba	18	11	16	11	26
Oleaceae	-	-	-	24	-
Parthenium hysterophorus	6	22	18	-	25
Pinus roxburghii	-	5	15	-	16
Poaceae	332	250	320	290	308
Polyad	1	-	4	-	-

Putranjiva roxburghii	-	-	-	-	5
Ricinus communis	32	5	7	42	45
Syzgium cumini	-	12	7	3	-
Terminalia sp.	6	-	-	25	6
Typha sp.	-	-	6	-	-
Urticaceae	7	48	44	27	53
Xanthium strumarium	4	2	-	-	15
Impatiens balsamina	2	-	-	-	-
Solanum sp.	15	-	-	-	-
Zea mays	-	-	6	-	-
Unidentified pollen	44	67	26	118	47

Comparison of present investigation (Burkard personal sampler) with the earlier pollen catches (Impaction air-sampler):

The present qualitative and quantitative dataset obtained on Burkard personal sampler could not be compared both with the incidences of pollen grains obtained earlier at BSIP, Lucknow due to completely changed methodology. However, qualitative comparison can be made. It was observed that a large number of pollen grains, given below, which were earlier registered, could not be detected in present investigation. *Artemisia vulgaris, Alternanthera* sp., *Alnus* sp., *Anethum graveolens, Albizzia lebbek, Betula* sp., *Cedrela toona, Croton bonplandianum, Citrus* sp., *Carica papaya, Coriandrum sativum, Chrozophora rottleri, Cycas circinalis, Dendrophthoe falcata, Ephedra* sp., *Emblica officinalis, Galphimia gracilis, Grevillea robusta, Jasminum* sp., *Ligustrum* sp., Lamiaceae, *Madhuca indica, Rumex dentatus, Polyalthia longifolia, Polygonum* sp., *Pithecolobium dulce, Prosopis juliflora, Pyrostegia venusta, Rauwolfia serpentina, Santalum album, Salmalia malabarica, Tribulus terrestris, Tamarindus inidca* and Trilete spores. Their absence may be accounted either due to a short exposure period of slide i.e. ten minutes weekly only or to the absence of such vegetation in vicinity. The meteorological conditions prevailing at that time may probably have also governed the dispersal of pollen grains. On the contrary, a few pollen grains such as *Cassia* sp., *Bougainvillea spectabilis, Zea mays, Solanum* sp. and members of Oleaceae have been recorded in the present investigation only. Out of these, there is possibility that some of the taxa had earlier been grouped under Caesalpinoidae (*Cassia*), Compositae (*Bougainvillea spectabilis*) and Poaceae (*Zea mays*).

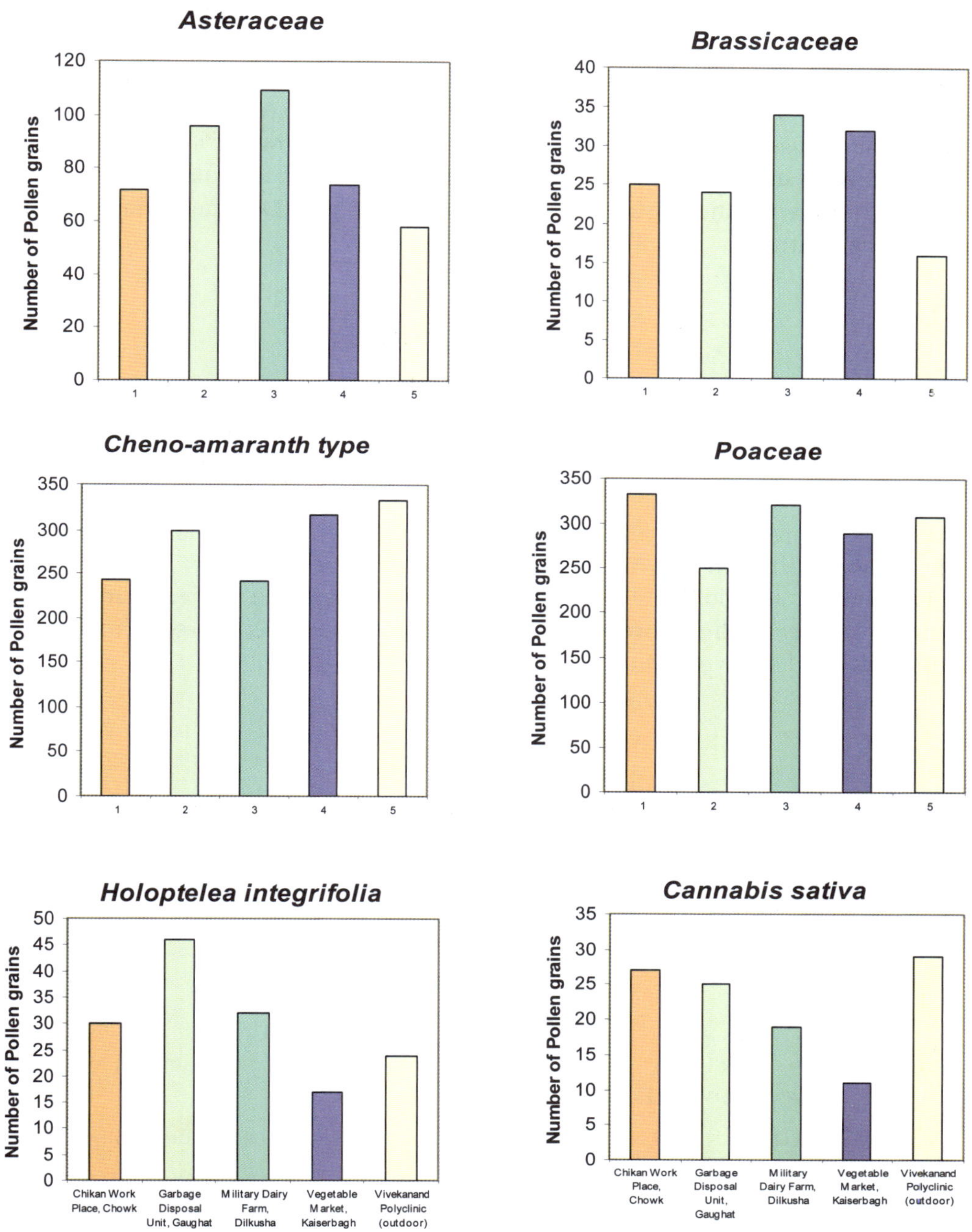

Figure 1. Number of dominant pollen grains of plant taxa

There is still some limitation in our study for example, a good amount of pollen grains remain unidentified which may either be due to introduction of new plants whose pollen morphology has not been studied so far or because of their long distance transport. Certain pollen grains continue to prevail irrespective of less exposure period or use of different sampler, which may be because of their wider distribution in surrounding vegetation and being high pollen producer (Khandelwal and Vishnu Mittre 1973). Our results demonstrate a decline in aerial pollen flora with the passage of time, which is result of rapid industrialization and urbanization.

CONCLUSION

Site to site variations in number of pollen grains have been recorded from different sites of Lucknow. So, the stress has been laid for periodical monitoring at different sites in order to provide specific zonal data to local clinicians/ allergists. Moreover, the results obtained by two different air samplers in Lucknow also posed difficulty in selecting the suitable air sampler for aerobiological investigation. In addition, in order to achieve the goal it is necessary to develop standardized methods/procedures, integrated national programmes, global aerobiological network and forecasting system.

ACKNOWLEDGEMENT

We thank the Ministry of Environment and Forest, New Delhi for funding AICP project entitled "Aeroallergens and Human Health: Aerobiological studies (42/14/94-RE)" during the tenure of which this work was done.

REFERENCES

Anonymous 1994-1998. AICP "Aeroallergens and Human Health: Aerobiological Studies (42/14/94-RE)". Ministry of Environment and Forest, New Delhi.

Chaturvedi, M., Datta, K. and Nair, P.K.K. 1987 88. A fresh survey of air borne pollen at Lucknow. *J. Palyn.* 23 24: 117 143.

Gopi. T.V., Prakash Kumar, Ravindran R. P. and Nair, P.K.K. 1990. Comparative analysis of airspora of two urban localities in Kerala. *Indian J. Aerobiol.* **3**: 39-44.

Gregory. P.H. 1961. The microbiology of the atmosphere Leonard Hill and Co. London, pp. 251.

Kapoor, S.L. 1962. On the botany of Lucknow district. *J. Bombay Nat. Hist. Soc.* 59(3): 863-896.

Khandelwal, A. and Vishnu Mittre 1973. Palynology of *Holoptelea integrifolia* Planch. *Palaeobotanist* 22(2): 127 132.

Khandelwal, A. 1988. Vertical variation of aerospora concentration at Lucknow, India. *Geophytology* 18(2): 173 185.

Khandelwal, A. 1989. Pollen and spore rain in Lucknow city during 1983 1986. *Proc. Int. Conf. on Biodeterioration of Cultural Property*: 20 25.

Khandelwal, A. 1992. Aeromycological survey at Lucknow. *Geophytology* 21: 199 206.

Khandelwal, A. 2001. Survey of aerospora by Rotorod Sampler: Qualitative and Quantitative assessment. *Aerobiologia* 17: 77-83.

Khandelwal, A. 2002. Long term monitoring of air-borne pollen and fungal sporesand their allergenic significance. *Palaeobotanist* 51: 153-159

Lakhanpal, R.N. and Nair, P.K.K. 1958. Survey of atmospheric pollen at Lucknow. *Jour. Sci. Indust. Res.* 17(1): 80 84.

Lacey, M.E. 1962. Summer air spora of two contrasting adjacent rural sites. *J. Gen. Microbiol.* 29: 485-501

Lewis, W.H., Vinay, P. and Zenger, V.E. 1984. Airborne and allergic pollen. *J. Allergy Clin. Immunol.* 51: 187 188.

Mari Bhat, M. and Rajasab, A.H. 1985. Incidence of airborne pollen at two different locations at Gulbarga during 1984-85. *Indian J. Med. Res.* 82: 346-352.

Oomachan, M., Misha, R.P. and Khare, N. 1988. Atmospheric pollen flora from two different sites at Jabalpur – a preliminary report. *Ind. J. Aerobiol.* 1: 66-70.

Smart, I. J. and Knox, R.B. 1979. Aerobiology of grass pollen in the city atmosphere of Melbourne: Quantitative analysis of seasonal and annual changes. *Aust. J. Bot.* 27: 317-331.

Shivpuri, D.N. 1980. Clinically important pollen, fungal, insect allergens for nasobronchial allergy patients in India. *Asp. Allergy. Appl. Immunol.* 13: 19 23.

Vishnu-Mittre and Khandelwal, A. 1973. Air borne pollen grains and fungal spores at Lucknow during 1969 70. *Palaeobotanist* 22(3): 177 185.

Advances in Pollen Spore Research Vol. XXVIII (2010) : 51-56
Editor : Dr. A. J. Solomon Raju
Today & Tomorrow's Printers and Publishers, New Delhi - 110 002 (India)

MEGASCOLIA FLAVIFRONS – AN UNUSUAL POLLINATOR OF ONION FLOWERS

D. P. Abrol

Division of Entomology,

Sher-e- Kashmir University of Agricultural Sciences & Technology,

Faculty of Agriculture,

Jammu – 18000, Jammu and Kashmir

Onion (*Allium cepa* L.) is a highly cross-pollinated crop depending upon cross-pollinating insects for any significant yield in seed production (Singh and Dharmwal 1970; Sihag 1986; Free 1993; Abrol 1997). The crop is protandrous with anthers shedding pollen before the stigma of the flower becomes receptive. Evidently, self pollination is not possible which necessitates the services of cross-pollinating insects for pollen transfer (Free 1993). A large number of insect species have been reported visiting onion flowers (Jabonski *et al.* 1982). Among all these insects, honeybees have, however, been reported as the most efficient and major pollinators of this crop because of their specific instinctive behavioural features for pollen and nectar collection and efficiency in pollen transfer (Free 1993).

Megascolia flavifrons

***Megascolia flavifrons* working on onion flowers**

The studies were made during May-June 2001 and 2002 at Shalimar Kashmir, India to record the insects visiting onion flowers. Among several insects frequenting onion flowers, honey bees, *Apis mellifera* L. *A.cerana* F. carpenter bee Xylocopa *valga* Gerstacker, some flies and wasps were the most important flower visitors. Also, a scolid wasp was collected visiting onion flowers. This wasp was found regularly visiting the onion flowers throughout the day and entire flowering season for nectar collection. Though it did not collect pollen deliberately but its behavior pattern was most suited for pollination of onion flowers as its body lay across many flowers on the umbels. It has been identified as *Megascolia flavifrons* Fabricius (Hymenoptera: Scoliidae) by the Commonwealth Institute of Entomology, London. It is large hairy with a yellow band on the abdomen. These wasps are usually parasites of the larvae of scarabaeid beetles (Borror and Delong 1963**)**. The adults are commonly found on flowers. The females burrow into the ground to locate a host; they find a grub, sting it, parasitize and then burrow deeper into the soil and construct a cell around a grub. Many grubs may be stung without a wasp ovipositing; such grubs usually never recover. In this wasp, the mesosternum and metasternum together form a continuous plate divided by transverse suture and overlapping bases of the middle and hind coxae, the hind coxae are well separated; wing membrane extends beyond closed cells with fine longitudinal wrinkles the apex of the abdomen of male has three retractile spines; it is a large brightly coloured wasp.

Behavior pattern:

Its behaviour pattern was most suited for pollination of onion flowers as its body lay across many flowers in the umbels. Emergence of *Megascolia flavifrons* coincides with the blooming of onion flowers during May-June, when temperature ranged from 25 to 33^{0}C. It has been observed visiting onion flowers throughout the day with maximum numbers between 11^{00}-13^{00} hours on clear sunny days. Though a slow worker visiting only 1.62 umbels/min. as compared to honeybees *A. cerana* (2.38) *A. mellifera* (2.78) and carpenter bees *Xylocopa valga* (7.18) (n=20). But its behaviour on flowers is ideal for cross-pollination. While lying across umbels their bodies got heavily dusted with pollens when the next flower was approached with body usually lying across the anthers. This posture seems to be essential for cross-pollination. Further, the pollen adheres loosely on the underside of the abdomen especially between the constriction of thorax and abdomen and is easily brushed off during visit to flowers. The number of pollen grains as determined by the method described by Abrol (1996) revealed that it carried on an average (n=20) higher number of pollen grains (2.59^{10}) as compared to honey bees, *A. mellifera* (1.31^{10}); *A. cerana* (1.29^{10}) and carpenter bee *Xylocopa valga* (1.60[10]). Evidently, they are valuable agents affecting cross-pollination in onion flowers besides acting as valuable biocontrol agents. Perusal of literature reveals that this wasp has not been recorded from India or elsewhere working on onion flowers though some wasps belonging to family scoliidae such as *Scolia quadripustulata* Fabr, *Elis annulata* Fabr. and *E. thoracica* Fabr.has was recorded visiting cotton flowers for nectar from India (Lefroy,1984). Evidently, its association with onion flowers constitutes new record from India or abroad.

Impact on seed production:

The data presented in Table 1 show that umbels left for open pollination resulted in significantly higher seed set than those excluded from insect visits. The caged umbels had 9.8 per cent seed set whereas those left for open pollination resulted in 93.5 per cent. Number of seeds produced per umbel also varied significantly. Caged umbels produced 1.9 seeds/capsule as compared to 4.3 seeds/capsule in case of umbels left for open pollination. Similarly, seed weight/ umbel and weight of 1000-seed also varied significantly in both the treatments. Open pollinated plots produced significantly more seeds with heavier weights than those isolated from insect visits ($P < 0.01$).

Table 1. Effect of pollination treatments on yield components of *A. cepa*

Parameter	No pollination (Enclosed)	Insect pollination (Open pollination)	CD at 5%
Percent fruit set	9.8	93.5	10.2
	(3.13)	(9.66)	
Number of seeds/capsule	1.9	4.3	14.22
	(1.37)	(2.07)	
Seed weight/umbel (g)	0.30	3.6	7.45
	(0.54)	(1.89)	
Seed weight/1000-seed (g)	2.9	3.2	0.47
	(1.70)	(1.78)	

Figures in parentheses are square root transformed values.

The data presented in Table 1 show that umbels left for open pollination resulted in significantly higher seed set than those excluded from insect visits.

PLATE

(a)

(b)

(c)

(d)

(e)

Plate 1. Experimental view: (a) Showing a general view of experimental trial of strawberry, (b), Apis mellifera (c) A florea (d) A.cerana working on strawberry flowers and (e) *Caged pollination of* strawberry

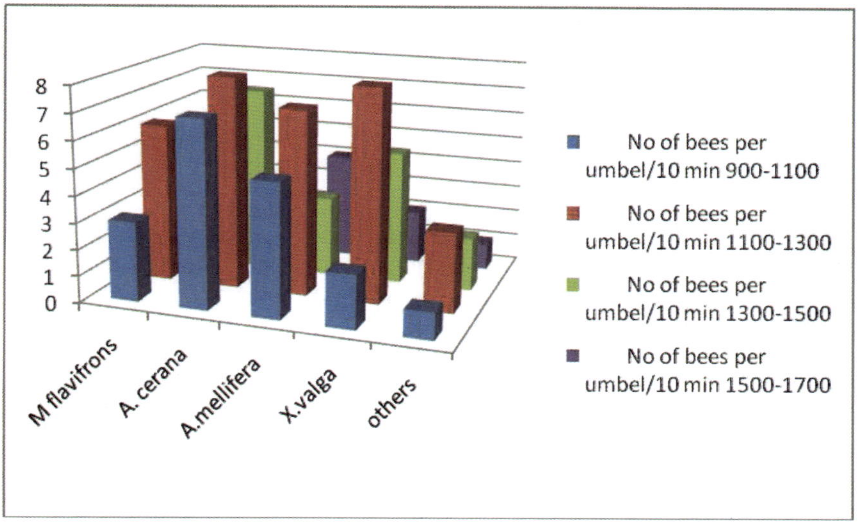

Figure 1. Diurnal activity pattern of bees visiting onion flowers

The caged umbels had 9.8 per cent seed set whereas those left for open pollination resulted in 93.5 per cent. Number of seeds produced per umbel also varied significantly. Caged umbels produced 1.9 seeds/capsule as compared to 4.3 seeds/ capsule in case of umbels left for open pollination. Similarly, seed weight/umbel and weight of 1000-seed also varied significantly in both the treatments. Open pollinated plots produced significantly more seeds with heavier weights than those isolated from insect visits ($P < 0.01$).

ACKNOWLEDGMENT

The author thanks Director International Institute of Entomology, London U.K. for identification of insect species.

REFERENCES

Abrol, D.P 1997. *Bees and Beekeeping in India*, Kalyani Publishers, Ludhiana, India, 450p

Abrol, D.P. 1996. Foraging activity of insect pollinators especially honeybees visiting apple flowers and their impact on fruit production. *Current research 25:196-199.*

Borror, D.J and Delong, D.M.1963. *An Introduction to the study of Insects*. Revised Edition, Holt, Rinehart and Winston, New York, Chicago, San Francisco, 819p.

Free, J. B. 1993. *Insect Pollination of Crops*.2nd Edition, Academic Press, London.

Jaboñski B., Skowronek J., Woyke H. W. and Doruchowski R. W. 1982. Biologia kwitnienia, nektarowanie, zapylanie i owocowanie mêskosterylnych linii cebuli (*Allim cepa* L.). *Pszczel. Zesz. Nauk.* 26: 57-104.

Lefroy, H. M. 1984. *Indian Insect Life: A manual of the insects of plains of tropical India.* Jagmender Book Agency, New Delhi, 786p.

Sihag, R. C. 1986. Insect pollination increases seed production in cruciferous and umbelliferous crops. *J. Apic. Res.* 25: 121-126.

Singh, J.P and Dharmwal, S.S. 1970. The role of honey bees in seed setting on onion at Pant Nagar. Dist. Nainital, Uttar Pradesh, India. *Indian Bee J.* 32: 23-27.

Advances in Pollen Spore Research Vol. XXVIII (2010) : 57-62
Editor : Dr. A. J. Solomon Raju
Today & Tomorrow's Printers and Publishers, New Delhi - 110 002 (India)

MELIPONICULTURE FOR YIELD ENHANCEMENT AND FOOD SAFETY IN KERALA

S. Devanesan, K.K. Shailaja and K.S. Premila

Kerala Agricultural University,
AICRP on Honeybees and Pollinators,
Department of Entomology,
College of Agriculture,
Vellayani- 695 522 Kerala

INTRODUCTION

Stingless bees are a class of hymenopteran insects under sub-family Meliponinae. These small-sized insects have a wide distribution in India. *Trigona iridipennis* is the common stingless bee found in India (Figure 1). In Kerala, it is reared as backyard beekeeping practice mainly for honey production. Meliponiculture is being popularized in the rural homesteads for poverty alleviation and additional income generation to women. The stingless bee honey is unique having great medicinal value which fetches high price up to Rs. 1000/- per kilogram. Concerted attempts are being made by All India Co-ordinated Research Project (AICRP) on Honey bees and Pollinators, Vellayani Centre of Kerala Agricultural University to improve the hiving, domestication, conservation and management of stingless bee with a view to enhance the number of colonies and there by meeting the current demand of stingless bee honey.

Hiving of feral colonies

The domicile of *T. iridipennis* is the basements of buildings, compound walls, tree trunks etc (Figure 2). Hiving of feral colonies from the above sites

Figure 1

Figure 2

lead to damage to the buildings and walls and hence the people hesitate to do the transfer of these colonies demolishing the structures into artificial hives. Hence different methods were tried at AICRP Centre, Vellayani in hiving feral colonies and a suitable method standardized. The mouth of an earthen pot has to be placed in front of the entrance of a feral colony and its rim is to be pasted to the wall using mud (Figure 2a,b). Proper support to the pot should be provided whenever required. An opening is to be given at the opposite side of the pot as bee entrance pasted with wax of stingless bee. The worker bees pass through the hole in the pot using it as new entrance. Gradually the bees realize that there is sufficient space inside the pot to store the brood and food material and hence they start construction of pollen and honey pots in the earthen pot. Eventually they build

Figure 2a

Figure 2b

the brood combs and start brood rearing inside it. The colony and pot should be left undisturbed for six months for the feral colony to settle. Later the pot with brood, pollen, honey, worker bees and a queen (Fig 3) can be detached and shifted to a suitable site to establish as a new colony.

Figure 3

Domestication of stingless bee

Stingless bees are domesticated in earthen pots, bamboo bits and wooden boxes (Figure 4a,b). Different types of hives *viz.,* wooden box, earthen pot, bamboo bit with volume of 1500cc, 2250cc, 3000cc, and 3750cc were tried for hiving stingless bee of which 1500cc capacity showed better brood development and honey storage than hives with different capacities (Figure 5).

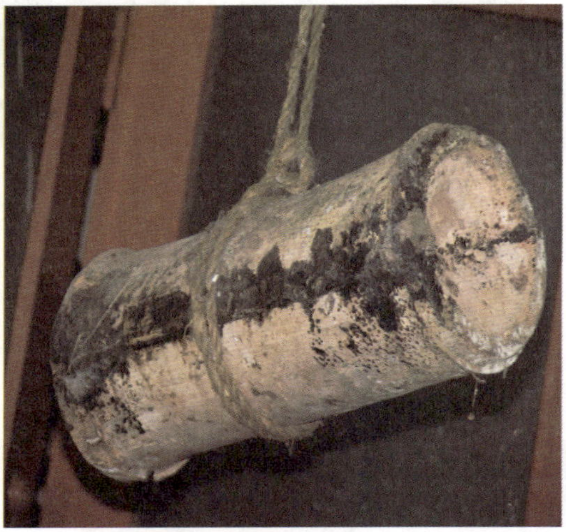

Figure 4

Figure

Figure 5

Stingless bee honey

The small amount of honey produced per hive is of great demand and fetches comparatively higher price due to its medicinal value. It is more acidic than commercial *Apis* honey but it has unique taste and medicinal properties (Figure 6).

Figure 6

The studies conducted at Vellayani centre indicated that the *Trigona* honey from different tracts of Kerala varied widely in colour from light yellow to dark amber colour with all intermediate shades. This is related to the varied flora in different locations. It also has specific flavour and aroma. Light coloured honey is mild in flavour and a darker honey has a more pronounced flavour. The percentage of moisture varied from 18.96 to 20.48% which was significantly lower than *Apis spp,* where the moisture percentage ranged from 19-23. The pH ranged from 3.76-4.4, which is lower than *Apis* honey. Stingless bee honeys were mostly multifloral. Hence variation in nectar quality occurred which influence honey characters like colour and acidity. The major sugars in stingless honey are glucose and fructose, which account to 37.1 and 41.6% respectively. Sucrose content varied from 1.2 – 1.8%. All the honey samples contain lesser amount of glucose. Mean per centage of total reducing sugars ranged from (77.1-69.85) and that of glucose content ranged from 37.36 -32.29. Percentage of fructose ranged from 39.85-37.57 in stingless bee honey. Specific gravity and refractive index are important properties. The specific gravity of *Trigona* honey ranged from 1.350-1.377. The refractive index was in the range of 1.4891-1.4855.

The stingless bee collects nectar and pollen from many medicinal plants with small flowers which are not visited by *Apis* species. This honey has great healing properties and effective for Asthma and allergy problems. It is beneficial in reducing the heat of the body, inflammation, bleeding in the throat, impurities of the blood, boils etc. It is used in many medicines in Indian System of Medicine, the Ayurveda for many diseases even for cancer.

Pollination service

Besides being a honey producer, stingless bees possess many characteristics that enhance their importance as crop pollinator both in wild and agricultural situations. The stingless bees visit most of the horticultural and agricultural crops for its pollen and nectar resulting in the pollination and yield enhancement. Augmentation, conservation, popularization and management of *T. iridipennis* should be intensified for ensuring sustainable agriculture and the conservation of biological diversity resulting in food safety.

Advances in Pollen Spore Research Vol. XXVIII (2010) : 63-70
Editor : Dr. A. J. Solomon Raju
Today & Tomorrow's Printers and Publishers, New Delhi - 110 002 (India)

SPORULATION IN MARINE BROWN ALGAE OF INDIAN COAST: A REVIEW

G. M. Narasimha Rao and G. Subba Rangaiah

Department of Botany

Andhra University

Visakhapatnam-530 003

ABSTRACT

Seaweeds are economically important for the sources of food, fodder, fertilizer and extraction of hydrocolloids, important bioactive compounds and biogas. In India, many seaweed based industries were shutdown due the shortage of the raw material. Brown algae commonly called as kelps, occur certain months of the year only. Cultivation of these seaweeds would generate the sufficient raw material for continuous running of the industries. Spores play an important role in the Mariculture. Studies relating to sporulation will be useful for generating the plant populations in the natural habitats. In the present paper data was summarized on the seasonal spore shedding, effect of environmental factors on spore shedding and diurnal periodicity on spore shedding of some marine brown algae such as *Sargassum swartzii, Sargassum wightii, Sargassum vulgare, Sargassum polycystum, Sargassum ilicifolium, Padina tetrastromatica, Padina boergesenii, Turbinaria decurrens, Turbinaria ornata, Cystoseira, Ectocarpus mitchellae, Rosenvingea nhatrangensis* and *Dictyota dichotoma* occur along the east coast and west coasts of India.

Key words: Marine brown algae, spore shedding, Indian waters

INTRODUCTION

In continuation with our earlier paper on red algal spore shedding (Narasimha Rao and Subba Rangaiah 2009), present paper deals with spore shedding in brown algae of the Indian coast. Spores are single celled reproductive bodies and are capable of growing into new organism. Brown algae include some of the largest and the most complex marine algae, the kelps, which grow up to 60 meters and forms dense underwater forests in the temperate coasts. In India different brown algae occurs on the infra littoral fringe region of the intertidal rocky surfaces and rock pools along the east and west coast regions. Brown algae (Phaeophyceae) play an important role for food, fodder and extraction of bioactive compounds. Brown algae provide natural sources for the manufacture of chemicals called alginates that are used as thickening agents and stabilizers in the industrial preparation of foods and pharmaceuticals. Biomass from the natural habitats is depleting due to various environmental factors. In these circumstances availability of raw material is one of the limiting factors for running the industry throughout the year. Therefore, cultivation of economically important seaweeds is very essential for generating the sufficient raw material for the requirement of the industries. Besides the vegetative propagation, spores play a vital role in commercial cultivation of seaweeds.

SEASONAL SPORE SHEDDING

Various authors (Chauhan and Krishna Murthy 1966; 1967; and 1971; Chennubhotla et al. 1978; Ganesan et al, 2000; Kaliaperumal and Umamaheswara Rao 1977; Kaliaperumal et al. 1977; Mairh 1968; Narasimha Rao 1989a;1989b;1991;1995; Narasimha Rao and Subba Rangaiah 1991; AppaRao 1995; 1998; Umamaheswara Rao and Kalimuthu 1972; Umamaheswara Rao and Kaliaperumal 1976) have studied the growth and seasonal spore shedding in marine brown algae such as *Sargassum swartzii, S. wightii, S. vulgare, S. polycystum, S. ilicifolium, Padina tetrastromatica, Padina boergesenii, Turbinaria decurrens, Turbinaria ornata, Cystoseira, Ectocarpus mitchellae, Rosenvingea nhatrangensis* occurring along the east coast and west coast regions of India. Most of the brown algal members are seasonal plants which grow in certain months of the year. During these periods plants produce tetraspores/oospores/plurispores and liberate in certain months of the year. Narasimha Rao (1989a; 1991) reported that the seasonal growth and seasonal spore shedding of the *Ectocarpus mitchellae* occurs along the intertidal rocky surfaces of the Visakhapatnam coast. Seasonal data collected on *Ectocarpus mitchellae* reveals that small plants occur in November and plants with maximum length in the month of January. Again from February onwards height of the plants decreases and in June this filamentous brown alga disappears from the study area. This seasonal growth data correlates the seasonal spore shedding and peak spore shedding in *Ectocarpus mitchellae* at Visakhapatnam coast. Liberation of plurispores from plurisporangia was observed in the month of November onwards. The output of plurispores increased from November to January and maximum outputs ranged between 132X10u and 126X10u spores in 1984-85 and 1985-86 respectively. Plurispores output decreased gradually from February and lowest value was obtained in the month of May. After May plants were not observed in the natural habitats of the Visakhapatnam coast. Maximum and minimum shedding of plurispores was coinciding with seasonal growth in this brown alga. Narasimha Rao (1995) studied the seasonal spore shedding in *Rosenvingea nhatrangensis* growing in the Visakhapatnam coast. Plants were observed from the month of November to March and no plants were reported in the month of April. Spore shedding was reported in *Rosenvingea nhatrangensis* from November onwards with maximum number of spore in December and January (7, 84,972 and 6, 92,846 spores/g.f.wt./day respectively). From February onwards spore shedding decreased with a minimum number in the month of March (3, 24,624 spores/g.f.wt./day). Appa Rao (1998) studied the seasonal oospore shedding in *Sargassum ilicifolium* and *Sargassum vulgare* growing at Visakhapatnam coast and observed the oospore shedding from October to April with maximum shedding in the months of December and February. The oospore output in *S. vulgare* ranged between 29,220 and 3, 05,983 and in *S.ilicifolium* it was 3,633 and 40,513. In these two species oospore liberation was noticed from day one to day twelve from the receptacles. Ganesan et al. (2000) collected the information on seasonal variation in growth and spore production of *Padina boergesenii* and *Padina tetrastromatica* in the Mandapam region, south east coast of India. Tetraspore output in *Padina boergesenii* showed the distinct seasonal periodicity with two peaks - one peak in December/January 1993 (19,260 spores/g.f.wt./day and 6300 spores/g.f.wt./day) and another peak in August 1994 (32,261 spores/g.f.wt./day).

EFFECT OF ENVIRONMENTAL FACTORS ON SPORE SHEDDING

Data on the effects of environmental factors such as exposed to air (desiccation) salinity, light intensity and different industrial effluents on the liberation of various spores from the marine brown algae of Indian waters was collected by several authors (Umamaheswara Rao and Sanjeeva Reddy 1982; Narasimha Rao 1989a; 1989b; Narasimha Rao and Subba Rangaiah 1991; Appa Rao 1995; 1998; Premila and Umamaheswara Rao 1997 and 1998; Murthy and Umamaheswara Rao 2003). In the experiments conducted by Umamaheswara Rao and Sanjeeva Reddy (1982) on tetraspore shedding in *Dictyota dichotoma* at Visakhapatnam coast revealed that the spore output was decreased with increase in the exposure of fronds to air and maximum out put was obtained under submerged conditions of the fronds. Maximum shedding of tetraspores was observed in the salinities between 25 and 35‰. Minimum output was noticed in lower and higher concentrations of salinity. Similarly maximum number of tetraspores was liberated between temperatures of 25 to 30p C. Narasimha Rao (1989b) collected the information on effect of environmental factors on spore shedding from *Rosenvingea nhatrangensis* present at Visakhapatnam coast. Maximum number of plurispore out put was recorded when fronds were in submerged condition. Spore out put was decreased with increasing time in the exposure to air and with minimum spores at 75 minutes of exposure. No spore shedding was reported beyond 75 minutes of exposure. Maximum plurispore out put was observed at 30‰ salinity. Spore out put was decreased from 30‰ salinity onwards and minimum at 50‰ salinity. Experiments conducted with different light intensities on plurispore shedding reveals that 1500 lux light intensity is favorable for maximum liberation of spores in *R. nhatrangensis*. Minimum number of plurispores shedding was reported at 500 and 2500 lux light intensities in this alga. Narasimha Rao and Subba Rangaiah (1991) studied the effect of environmental factors on the spore shedding in *Ectocarpus mitchellae* growing at Visakhapatnam coast. In their study, they reported that the maximum number of plurispores was liberated when fronds were in submerged condition. Minimum number of spores was reported at 60 minutes exposure and spore shedding completely ceased when fronds were exposed to 80 minutes and onwards. In the experiments conducted with different salinities, maximum plurispore out put was noticed at 30‰ salinity. No plurispore shedding was observed at 10 and 60‰ salinities. The effect of different light intensities is ranging from 'o' (dark) to 97 μE m^{2} s^{1} on spore shedding indicates that maximum number of plurispores was observed at 27 μE m^{2} s^{1}. Plurispore out put decreased from 27 μE m^{2} s^{1} onwards and there was no shedding at 79 μE m^{2} s^{1}. Appa Rao (1995) studied the effect of environmental factors on tetraspore shedding in *Padina tetrastromatica* growing at Visakhapatnam coast. Maximum number of tetraspores were released when fronds in submerged state and the spore out put was decreased slowly as the period of exposure increased and the liberation of spores was seen up to 6 hours in this alga. Optimum tetraspore shedding was noticed at 30‰ salinity and 500 lux light intensity. Working with the species of *Sargassum ilicifolium* and *Sargassum vulgare*, Appa Rao (1998) observed that maximum number of oospores were liberated when receptacles in the submerged condition. Among the various salinities tested, 20 and 30 ‰ salinities are found to be favorable for maximum liberation of oospores. And peak shedding of oospores was found at 25 p C temperature.

Premila and Umamaheswara Rao (1998) studied the effect of petroleum refinery and fertilizer factory effluents on marine brown algae such as *Padina tetrastromatica* and *Dictyota dichotoma.* Spore shedding was observed in all concentration of V.R.F. (Visakha petroleum Refinery) from 10 to 100%. In *Padina tetrastromatica* and *Dictyota dichotoma*, spore out put was decreased from 20% concentration. Minimum number of tetraspores was liberated from 30% onwards. No spore discharge was recorded beyond 100%. In *Dictyota dichotoma* spore shedding was also reported in all concentration of C.F.E (Coromandel factory effluent). Maximum number of spores was liberated at control and 25% concentration of effluent. Beyond, 25% concentration no spore discharge was observed. Murthy and Umamaheswara Rao (2003) collected the information on effect of industrial wastes on the growth and reproductive stages of macro algae of Visakhapatnam coastline, east coast of India. Studies were carried out on spore shedding of *Padina tetrastromatica* and *Sargassum polycystum* in different concentration of industrial wastes. The effect of only two wastes (Hindustan zinc effluent and Alum factory effluent) on tetraspore shedding of Padina indicates that median effective concentrations (24 Ec 50 values) of all the wastes, estimated for the spore liberation in *P. tetrastromatica*. Spore out put decreased rapidly from 2.0 -10.5 %in different effluents tested. There was no spore discharge after 60% in *P. tetrastromatica*.

DIURNAL PERIODICITY IN LIBERATION OF SPORES

Circadian rhythms are endogenously and exogenously generated, and can play a vital role in liberation of spores. The impact of diurnal rhythm on spore shedding of marine algae was studied by several authors in different geographical regions of the globe. Diurnal periodicity on spore shedding in brown algae is very limited (Umamaheswara Rao and Sanjeeva Reddy 1982; Narasimha Rao 1989a; 1995; Narasimha Rao and Subba Rangiah 1991; Appa Rao 1995 and 1998; Ganesan et. al. 2000). Umamaheswara Rao and Sanjeeva Reddy (1982) conducted the experiments with *Padina tetrastromatica* on diurnal periodicity of tetraspore shedding and observed that maximum number of spores were liberated between 0600 and 1000 hours. Umamaheswara Rao (1990) summarized the work done on the oospore shedding from *Sargassum vulgare* and *Sargassum ilicifolium* growing at Mandapam coast and Visakhapatnam coast and reported that maximum number of oospores from the above two species were released between 0200 and 0600 hours. Narasimha Rao and Subba Rangaiah (1991) studied the diurnal periodicity of spore shedding in *Ectocarpus mitchellae* at Visakhapatnam coast and observed that no plurisopre output in the early morning hours, maximum number of spore out put was observed between 1000 and 1400 hours and less number was reported in the next four hours period. Narasimha Rao (1995) studied the diurnal periodicity in spore liberation in *Rosenvingea nhatrangensis* from Visakhapatnam coast and observed that maximum spore shedding was noticed in between 0200 and 0600 hours and minimum spore shedding was recorded between 1000 and 1400 hours. No spore shedding was reported between 1400 and 2200 hours. Experiments conducted with *Padina tetrastromatica* (Appa Rao 1995) on diurnal periodicity of spore shedding, observed that maximum number of tetraspores were released from the fertile fronds between 0600 and 1000hours. Appa Rao (1998) studied the diurnal periodicity on oospore shedding in *Sargassum ilicifolium* and *Sargassum vulgare* growing at Visakhapatnam coast and reported that optimum oospore shedding was observed between 0200 and 0600 hours. Ganesan *et al.* (2000) working with two brown algal forms such as *Padina tetrastromatica* and *Padina boergesenii* observed that diurnal periodicity in the liberation of tetraspores shows increase in spore output during dark period from 1800 hours onwards and maximum number of spores were released between 2200 and 0200 hours.

CONCLUSIONS

Species of the marine brown algae are seasonal plants which occur from the month of November to April on the rocky surfaces of the inter-tidal regions. Seasonal spore output was higher during the months of December and January and spore shedding coincided with the maximum growth periods of the algae. More number of spores was liberated from the brown algae than the red algal members. Regarding the effect of environmental factors on spore shedding, it was observed that salinity in between 20 and 30‰, low light intensity; submerged conditions of the fronds are favorable factors for the maximum liberation of spores. The diurnal periodicities exhibited by different algae also shows that different algae might have selected different timings for their maximum liberation for better settlement and also avoid feeding by the different animals in the marine environment and nature itself might have set the timings for better settlement and to avoid feeding of spores.

REFERENCES

Appa Rao, D. 1995. Studies of spore shedding in *Padina tetrastromatica* Hauck (Phaeophyceae) of the Visakhapatnam coast. M.Phil. Dissertation, Andhra University, Waltair.

Appa Rao, D. 1998. Studies on the ecology of some species of *Sargassum* (Fucales, Phaeophyceae) along the Visakhapatnam coast. (India). Ph.D. Thesis, Andhra University, Visakhapatnam.

Chauhan, V. D. and Krishna Murthy, V. 1966. Observations on the output of oospores their liberation, viability and germination in *Sargassum swartzii* (Turn.) C. Ag. Salt Research and Industry 3: 11.

Chauhan, V. D. and Krishna Murthy, V. 1967. Observations on the output of oospores their liberation, viability and germination in *Sargassum swartzii* (Turn.) C. Ag. In: V. Krishna Murthy (ed), Proc. of the seminar on sea, salt and plants. Bhavnagar, India. pp. 197-201.

Chauhan, V. D. and Krishna Murthy, V. 1971. Ecology and seasonal succession of *Sargassum swartzii* (Turn) C. Ag. in Indian waters. Phykos 10: 1-11.

Chennubhotla. V. S. K., Krishna Murthy, V., Kaliaperumal, N. and Kalimuthu, S. 1978. Seasonal changes in growth, fruiting cycle and oospore out put in *Turbinaria conoides* (J. Agardh) Kutzing. Bot. Mar. 21: 67-69.

Ganesan, M., Mairh, O. P and Subba Rao, P.V. 2000. Seasonal variations in growth and spore production of marine brown algae *Padina boergesenii* and *P. tetrastromatica* (Dictyotales/ Phaeophyta) in the Mandapam region, south east cast of India. Indian J. Mar. Sci. 29: 253-257.

Kaliaperumal, N. and Umamaheswara Rao, M. 1977. Growth, fruiting cycle and oospore output in *Turbinaria decurrens* Bory. Indian J. Fisheries 22: 225-230.

Kaliaperumal, N., Chennubhotla. V. S. K. and Kalimuthu, S. 1977. Growth, reproduction and liberation of oospores *Turbinaria ornata* (Turner) J. Agardh. Indian J. Mar. Sci. 6:178-179.

Mairh, O.P. 1968. Observations on the germination of oospores and growth of germlings in *Cystoseira*. J. Ind. Bot. Soc. 22: 256-263.

Murthy, S. B. K. and Umamaheswara Rao, M. 2003. Effects of industrial wastes on the growth and reproductive stages of macro algae of Visakhapatnam Coastline, east coast of India. Indian J. Mar. Sci. 32: 57-66.

Narasimha Rao, G. M. 1989a. Ecological studies on some estuarine and marine algae. Ph.D. Thesis, Andhra University, Waltair.

Narasimha Rao, G. M. 1989b. Effect of environmental factors on spore shedding from *Rosenvingea nhatrangensis* Dawson. Mahasagar 22: 135-137.

Narasimha Rao, G. M.1991. Seasonal growth and spore shedding in *Ectocarpus mitchellae* (Harv.) Hamel. Mahasagar 24: 59-62.

Narasimha Rao, G. M.1995. Seasonal variation and diurnal periodicity in spore liberation from *Rosenvingea nhatrangensis* Dawson. Phykos 34: 145-148.

Narasimha Rao, G. M. and Subba Rangaiah, G. 1991. Control of spore shedding from some marine algae of the Visakhapatnam coast, India. Br. Phycol. J. 26: 353-360.

Narasimha Rao, G. M. and Subba Rangaiah, G. 2009. Spore shedding in marine red algae of the Indian coast: A review. Advances in pollen spore Research. XXVII: 43-53.

Premila, V. E. and Umamaheswara Rao, M. 1997. Effect of crude oil on the growth and reproduction of some benthic marine algae of Visakhapatnam coastline. Indian J. Mar. Sci. 26: 195-200.

Premila, V. E. and Umamaheswara Rao, M. 1998. Effect of Refinery and fertilizer factory effluents on Marine algae of Visakhapatnam coast. Phykos 37: 41-49.

Umamaheswara Rao, M. 1990. Autecological and Eco-physiological studies on marine algae of Visakhapatnam and Mandapam coast. Perspectives in Phycology (M.O.P. Iyangar Centinary celebration Volume) Ed. V.N.Raja Rao. pp. 323-335.

Umamaheswara Rao, M. and Kalimuthu, S. 1972. Changes in mannitol and Alginic acid contents of *Turbinaria ornata* (Turner) J. Agardh in relation to growth and fruiting. Bot. Mar. 15: 57-59.

Umamaheswara Rao, M. and Kaliaperumal, N. 1976. Some observations on the liberation and viability of oospores in *Sargassum wightii* (Greville). J. Agardh. Ind. J. Fish. 24. 232-235.

Umamaheswara Rao, M. and Sanjeeva Reddy, R.B. 1982. Influence of desiccation, salinity and temperature on the liberation and germination of *Dictyota dichotoma*. Seaweed Res. Utiln. 6:5-9.

Advances in Pollen Spore Research Vol. XXVIII (2010) : 71-82
Editor : Dr. A. J. Solomon Raju
Today & Tomorrow's Printers and Publishers, New Delhi - 110 002 (India)

APOMIXIS IN AN ENDANGERED ENDEMIC TREE SPECIES *POECILONEURON PAUCIFLORUM* BEDD. (CLUSIACEAE) FROM THE WESTERN GHATS, SOUTH INDIA

Y. Justin Koilpillai and S. Wilson

Department of Biotechnology, Sathyabama University, Chennai 600 119, Tamil Nadu

ABSTRACT

Poeciloneuron pauciflorum Bedd. (Clusiaceae) is an endangered, medium sized evergreen tree species endemic to the Agasthiamalai of Southern Western Ghats. Based on ecology, morphology, seed biology, cytology, tissue culture studies and molecular studies, it is concluded that this endemic tree species is an apomictic species. The absence of vegetative apomixis has been confirmed through tissue culture and field experimental studies. The production of normal seeds, presence of multiple stigmas, variation in number of petals, presence of polyembryony in seeds and the operation of amitotic division in pollen mother cell indicate the operation of apomixis (agamospermy) in *Poeciloneuron pauciflorum*.

Keywords: *Poeciloneuron pauciflorum*, apomixis, endemic, cytology, RAPD.

INTRODUCTION

The Western Ghats in India is one of the twelve important hotspots of the world with rich biodiversity. In general there are about 857 species of trees in various types of forests on the Western Ghats, of these 540 species are endemic to the Western Ghats (Ramesh and Pascal 1997). Endemic species belonging to the genera such as *Poeciloneuron, Calophyllum* and *Hopea* are dominant species in various types of forests on the Western Ghats. Out of 540 endemic species 137 species have been listed as rare and endangered ones. The endemic genus *Poeciloneuron* is represented by two species namely *Poeciloneuron indicum* Bedd. and *P. pauciflorum* Bedd. on the Western Ghats. The species *Poeciloneuron pauciflorum* is a rare one and it has been reported from the extreme southern tip of the Western Ghats (Tirunelveli Hills). It is used for walking sticks and building construction (Gamble 1921).

Beddome (1871) first described this species based on the specimen from Madras Herbarium. After Beddome, Dyer (1874) reported this species after 3 years. Later, Dunn (1957) reported it after long a gap of 73 years. Subsequently it was rediscovered by Vajravelu and Daniel (1983), Ramamurthy (1983) and Ahmedulla and Nayar (1990). All the above records are mainly from the specific region i.e., Tirunelveli Hills and Travancore region of Agasthiamalai of Southern Western Ghats. Thus, it seems to be a very rare species occurring in very few localities.

Nayar and Sastry (1990) have recommended this rare species to be conserved. According to them loss of natural habitat compounded by over exploitation or the classic "Over-kill" are primarily responsible for its depletion in the wild. This species, as one of the two constituent members of an endemic genus confined to southern peninsular India, should be conserved immediately. In a right way this species is one among the four species selected for the conservation program under "National Species Recovery Program" sponsored by the Department of Biotechnology, Ministry of Science and Technology, New Delhi (2006). In any conservation program the details about the target species regarding its ecology, distribution, reproductive behaviour etc are necessary. In order to know whether there is successful normal sexual reproduction i.e. Amphimixis or any other mode of reproduction i.e. Apomixis in this rare and endemic species, a detailed study was carried out both in the field and in the laboratory.

MATERIALS AND METHODS

Field survey was carried out on Agasthiamalai region, Western Ghats, from 2002-2005 to know the present status of the population of this species along with other details. During this period the phenology was also studied. Macro and micro morphological studies were carried out both in the field and in the laboratory by using hand lens and stereomicroscope. Cytological study was carried out by using young flower buds fixed in Carnoys Fluid (Mixture of Absolute Alcohol:

Acetic Acid: Chloroform 6: 3: 1). Young anthers were squashed in Acetocarmine stain to observe various stages of meiosis in Pollen Mother Cells. Various stages of meiosis were photographed by using Motic Microscope. For *in vitro* studies explants such as node, internode, shoot tip and embryo were cultured in MS basal. To study the genetical variation, RAPD analysis was carried out by following the method of Doyle and Doyle (1987).

RESULTS AND DISCUSSION

The field survey on Agasthiamalai region, Western Ghats of South India shows that the species *Poeciloneuron pauciflorum* is growing on four different localities namely Mahendragiri (950m), Veyillar (950m), Inchikuzhi (650m) and Kannikatty (750) (Figure 1 a-g). In general the frequency of mature individuals is 5 –11 in a Quadrat (30 x 30m) and the frequency of saplings and seedlings is low in a Quadrat (10 x 10m) with 2-3 and 4-8 respectively. In all the localities the trees are growing along a river or stream in evergreen forests. The temperature in all the localities varies from 21-23° C and the humidity ranges from 75-85%. The diameter of the trunk varies from 27 to 73 cm. The height ranges from 7 to 11m. The emergence of leaf buds is seen in the last week of February and August. The flowering is seen in last week of March and October. The total period of flowering is between 35-40 days. The maximum anthesis is seen between 12 noon to 2 pm. The fruit setting starts in the last week of April and November.

This medium size riparian species has oppositely arranged, narrowly elliptic or lanceolate, 10 – 13 x 3 – 4 cm, glabrous leaves with entire margin and bluntly long acuminate apex. Flowers are white, solitary, 1-1.2 cm across, terminal or at each node in axils of fallen leaves. Sepals are four of unequal size. Petals are 5-8 arranged in imbricate manner. Stamens are 25 –31 in number with 4-5mm long anthers. Pollen grains are uniform, about 4250 per anther and are about 23 µm in diameter with smooth surface. Ovary is bicarpellary and bilocular with two ovules in each locule. There are two or three styles with acute stigma. Fruits are capsule, globose, 0.7 x 1.2 cm in size with persistent styles and sepals. Seeds are generally ovoid with two equal cotyledons and one embryo, rarely irregular in outline with three cotyledons and two embryos. The testa is hard membranous, striated and removable (Figure 2 a-m).

Meiotic behaviour was observed in Pollen Mother Cells in young anthers. There is no indication of regular pathway of meiosis, but as in regular meiosis four daughter cells are formed from each pollen mother cell. It is important to note that the chromosomal stage is very rare. Out of about 2000 pollen mother cells observed, only two were with chromosome stage (Figure 3g,h). All the other cells are with one, two or four nuclei with several intermediate stages of direct division of the nucleus by stretching of the entire nucleus without the formation of distinct individual chromosomes (Figure 3c-f). It is clear that the nucleus of each pollen mother cell undergoes two subsequent amitotic divisions resulting in four daughter cells with unreduced or with unbalanced chromosome number instead of four haploid daughter cells in normal meiosis. Although it is of abnormal pathway of cell division in pollen mother cells, there is no indication for the presence of three nuclei or more than four nuclei of different size. Usually in irregular meiosis presence of micronuclei in the daughter cells is very common (Manickam and Irudayaraj 1988). In the present case, there is no indication for the presence of any micronuclei at any stage. Pollen grains in mature anthers are uniform in size and shape.

In *in vitro* studies, the viability of both entire seeds and dissected embryos and the regenerative capacity of the nodal and shoot tip explants were tested. The viability of the seeds and the dissected embryos are same with 70% germination, which was also observed in conventional seed germination method. In contrast, there is no successful proliferation from the explants such as node and shoot tip. There is initiation (55%) of some shoots from the node but they remain as such without further growth and they ultimately withered. There is no indication for the improvement even with the addition of various growth hormones (BAP, NAA, 2IP, Kinetin & TDZ) with different concentrations and combinations. About 2% seeds produced twin seedlings in conventional method of seed germination in the green house, for which the seeds were selected randomly. But all the seeds produced single seedling under *in vitro* culture for which uniform sized seeds were selected.

Random Amplified Polymorphic DNA (RAPD) analysis on fourteen samples from two localities (Inchikuzhi and Kannikatty) with the calculation of matrix using 12 divers primers versus 14 accessions showed 85% similarity coefficient.

From the above observations it is clear that this species shows very narrow distribution growing in particular ecological niche (shaded stream bank or river bank) at medium altitude with low temperature (21-23°C) and high humidity (75-85%). This species struggles for its existence in nature due to the absence of normal and successful reproduction but not due to the human disturbance. They produce viable seeds. But the frequency is very low due to several reasons, which result in the production of low number of offspring.

The morphology of this species, particularly floral morphology, indicates the irregularity. The number of individual parts of the flower is not constant. The number of petals varies from four to nine and the number of anthers varies from 25-31. The embryo in a seed is also one or two in number. Styles and stigmas are two or three (Figure 2f-h). The presence of variable number of floral parts indicates the hybrid origin of this species, which is also supported by the absence of normal meiosis.

Among the floral parts, the morphology of the stigma is very peculiar in having acute apex, which is the rare type of stigma in angiosperms, and the most common type is the stigma with capitate head. The main function of the stigma is to provide the pollen with water necessary for their germination. In many plants it also supplies the necessary medium for pollen germination in the form of exudates. The stigmas, which secrete exudates, are called wet stigmas and those, which do not, are called dry stigmas (Bhojwani and Bhatnagar 1974).

The acute stigma in *Poeciloneuron pauciflorum* is smooth without any kind of hairs. It is not the suitable place for the germination of pollen grains after pollination. This impairment in the morphology of the stigma itself may be one of the reasons for the rarity of the species. In order to verify this reason, all the endemic tree species of Agasthiamalai were verified for the type of stigma with the detail about their distribution (Gopalan and Henry 2000; Mohanan and Sivadasan 2002). Out of 45 tree species, details about the type of stigma is available for 32 species. There are 12 species with capitate type of stigma. Of the twelve species, eight species are critically endangered, two species are endangered, one species is vulnerable and another species is probably extinct. There are seven species with acute and out of seven species three are endangered, three species are probably extinct and one species is critically endangered. Bifid stigma is found in three species, which are all critically endangered. Two species are with fusiform type of stigma and both are endangered species. There are nine other types of stigmas such as obscur, peltate, reniform, curved, flat, papillose, truncate, globose and subulate with single species in each type and all, except one (endangered), are critically endangered. Moreover, the extension of occurrence of species with acute stigma is very narrow (10 –20 km^2).

From the above details, it is surprising to see that, out of 32 endemic tree species with different kinds of stigmas, there are four species with "probably extinct" status. Out of four extinct status species there are three with acute type of stigma and one is with capitate stigma. From this data it is very clear that the species with acute stigma are usually rare and endangered but the species with other type of stigma may or may not be rare or endangered.

The presence of amitotic division in the place of meiosis in the anther itself is enough for the impairment of normal sexual reproduction i.e. amphimixis. The formation of seeds shows the presence of apomixis i.e. Agamospermy, where

Plate - 1

Figure 1: General views of different localities, habitat and population of *Poeciloneuron pauciflorum* on Agasthiyar Hills

a. A general view of Agasthiamalai and its environs; b. View of evergreen forest in Mahendragiri region; c. River bank in a forest at Kannikatty forest region; d. Veyillar river; e. An entire view of medium sized tree at river bank of Inchikuzhi; f. Tree with flushing in Kannikatty population; g. Group of samplings in between the forest region of Inchikuzhi and Kannnikatty

Plate - 4

Figure 2: Flowers, Fruits, seeds and seedlings of *Poeciloneuron pauciflorum*

a. A Flowering twig; b. Flowers with variable numbers (6, 8) of petals; c. Close up view of pistil with calyx; d. An individual petal; e. Anthers; f. Close up view of style and stigma with calyx; g. Pistil with two styles; h. Microscopic view of style with acute stigma; i. Fruit with persistent stigma; j. A twig with fruit; k. Mature seed; l. Mature fruits at dehiscence stage; m. Normal single and abnormal twin seedlings

Plate - 6

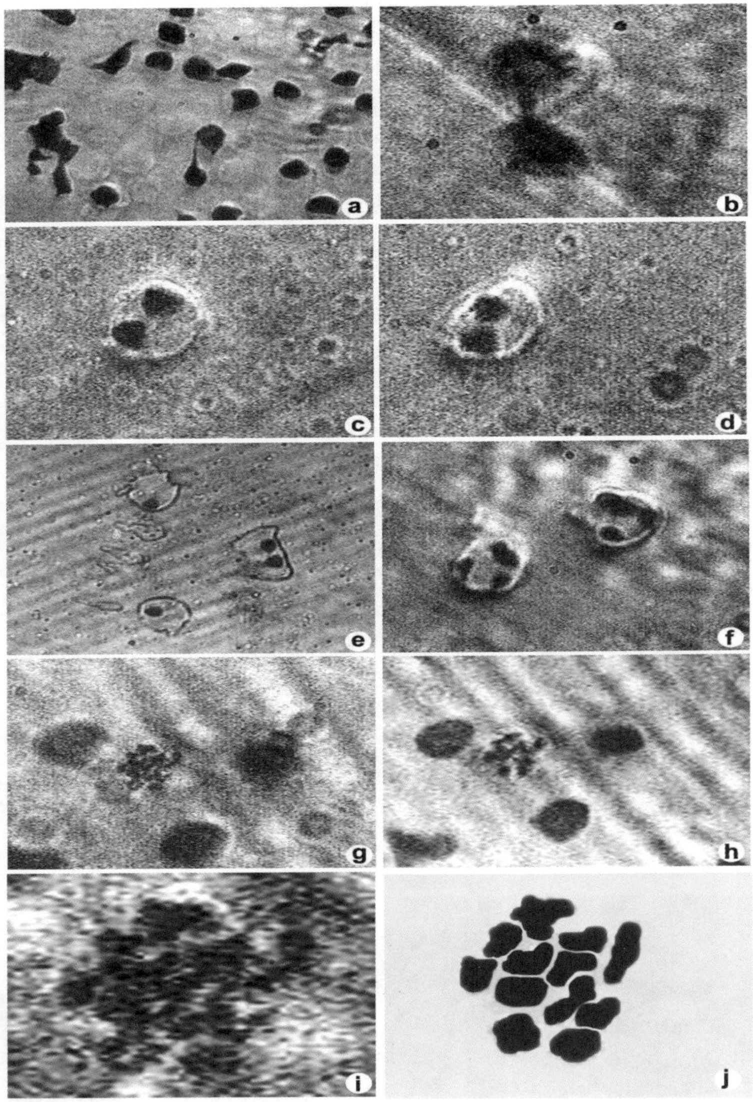

Figure 3: *Poeciloneuron pauciflorum*: Abnormal cell division in Pollen Mother Cells (PMC)

a. Amitotic division in several PMCs resulting in two daughter nuclei by nuclear stretching with chromosomal bridge in several cases; b. Amitotic division in single PMC in two daughter nuclei connected by chromosomal bridge; c. & d. Single PMCs with two daughter nuclei (Note angular inner end); e. PMCs with single or double nuclei; f. PMCs with four daughter nuclei after two amitotic divisions; g. & h. Two different PMCs at diakinetic stage of meiosis with approximately 12 bivalents; i. Enlarged view of figure g with 12 bivalents; j. Explanatory diagram for figures i. showing 12 bivalents.

reproduction through apomixis involves formation of seeds. In apomictic *Zephyranthes texana* Herb. the megasporocytes undergo entirely somatic divisions without meiotic chromosome pairing or contraction. At the metaphase of this apomeiotic division the chromosomes are mitotic and the resulting divisions do not reduce the chromosome number (Brown 1951). In the case of apomictic, *Elymus rectisetus* apomeiotic division is delayed relatively to meiotic division because of a lengthy period of megaspore mother cell vacuolation during nuclear stretching that occurs in the place of meiosis I. Amitosis is evident enough to possibly 5% of the megaspore mother cell during nuclear stretching. A subsequent mitotic division facultatively functions as meiosis II (Craw and Carman 1987). In the present case, the meiosis is replaced by amitosis in more than 98% pollen mother cells.

There are several criteria, which can be used as indication of apomixis. These criteria only suggest the presence of apomixis, which may later be confirmed by a combination of cytological, genetic and progeny tests. The criteria other than progeny tests are: 1. High seed set and/or uniform progeny in plants with irregular chromosome number or meiotic behaviour. 2. Multiple seedlings per seed, flowers with two or more fused ovaries, multiple stigmas and multiple ovules per floret (Gupta 1985). In *Poeciloneuron pauciflorum* considerable seed setting is seen. In the anthers regular meiosis was not observed, but it has been replaced by two amitotic divisions to form four daughter cells from single pollen mother cell (Figure 3 c-f)). RAPD analysis shows that the genetic variation in the population is very low with 85 % similarity. Flowers are with irregular number of petals, anthers, style and stigmas. Seeds are also with irregular number of cotyledons and embryos. About 2% seeds are with twin seedlings. The irregular shape and surface of several fruits may indicate the fusion of ovaries. Even the environment is also favourable for the apomitic mode of reproduction of this species, which usually grows along shaded river or stream. It has been proved that long days encouraged sexuality and continuous short days encouraged apomixis in *Dichanthium aristatum* (Knox and Heslop-Harrison 1963).

The causes of apomixis as indicated by Stebbins (1950) are: 1. Hybridization and allopolyploidy between the original sexual ancestors; 2. Hybridization between facultative apomicts and sexual species and 3. Chromosomal and genetic changes within the apomitic clones themselves. In the present case clear chromosome count was not possible due to the absence of any chromosomal stage in Pollen Mother Cells. Observation on two cells with chromosomal stage shows the presence of low number of chromosomes (Figure 3). There is no previous report on chromosome number for this species and other related species. *Garcinia mangostana* is a polyploid (2n=90), probably a tetraploid and would appear that it may be an allotetraploid derivative of *G. malaccensis*

(2n=42) and *G. hombroniana* (2n=48) (Richards 1990). Variable chromosome number of mangosteen has been reported, 2n= 56-76, 88-90, 96 and 120-130 (Verheij 1991).

Peter Stevens (Pers. Comm.) stated that "*Poeciloneuron pauciflorum* is a sister taxon to other two species of *Poeciloneuron* (one not published)" In molecular studies *Poeciloneuron* may be linked with *Kayea*, a very largely malesian genus that has one species, *K. stylosa* growing in Sri Lanka. They have similar stigmas. From the statement of Peter Stevens (Pers. Comm.), it is clear that the present species *P. pauciflorum* may be of hybrid origin and the parents are yet to be confirmed. One parent may be of the only related species of this genus *P. indicum* growing on the Western Ghats, Karnataka.

In conclusion, the indication for the hybrid origin of this species, replacement of meiosis by amitosis in anthers, presence of twin seedlings per seed, presence of multiple styles and stigmas, growing of the individuals in cool shade environment and the presence of genetically uniform individuals indicate the failure of normal amphimixis and operation apomixis (agamospermy) in *Poeciloneuron pauciflorum*. The absence of vegetative reproductive potentiality both *in vitro* and *in vivo* studies and the production of viable seeds even without regular meiosis indicate the presence of agamospermy in the rare endemic tree species *Poeciloneuron pauciflorum*.

Recently the past efforts, future strategies and potential application of apomixis research in India have been reviewed by Bhat *et al* (2005) and Kaushal *et al.* (Kaushal et al. 2005). The discovery of apomixis in forest trees or its transfer to, and utilization in these trees would have far-reaching potential in the improvement of economically important trees. The utilization of apomixis in cultivated trees presents a challenge for research teams to combine tree breeding, cytogenetics, cell culture and molecular biology techniques to make maximum use of apomixis in tree breeding. Apomixis is an attractive trait for the enhancement of crop species because it mediates the formation of large genetically uniform populations and perpetuates hybrid vigour through successive seed generations. The value of these opportunities will vary between crops and between production systems (Bhatt et al. 2005). Since *Poeciloneuron pauciflorum* is medium sized endemic tree with apomictic mode of reproduction, it will be of useful species for tree breeding.

Any effort to multiply through micropropagation by using vegetative parts of this species will be difficult due to the diploidal nature of this species, which are usually with low vegetative reproductive capacity when compared to polyploid species (Hagerup 1932). Such efforts are also temporary ones. In contrast, if this diploid hybrid species is made into a tetraploid one to get an auto-allotetraploid it will be of more successful method which will result in a normal sexual species with normal meiosis and high reproductive capacity. Ordinary breeding methods or cytogenetical methods or any other biotechnological methods may be followed to achieve this goal in order to conserve this rare endemic tree species.

ACKNOWLEDGEMENTS

Authors are thankful to Rev. Dr. A. Antonysamy S.J., Principal, St. Xavier's College (Autonomous) for his encouragements. Financial assistance from the Department of Biotechnology, New Delhi is also acknowledged. We are also thankful to Dr. Peter Stevens for his valuable suggestions and comments.

REFERENCES

Ramesh, B. R. and Pascal, J. P. 1997. Atlas of Endemics of the Western Ghats (India) Distribution of tree species in the evergreen and semi – evergreen forests. Institute Fracais de Pondichery, Pondichery.

Gamble, J. S. 1921. *Flora of the Presidency of Madras*, Vol. I. Bishen Singh Mahendrapal Singh, Dehradun, India.

Beddome, R.H. 1871. In: *Flora of Sylvaticum* p. 93.

Dyer, G. 1874. In: *Flora of British India*. 1: 278.

Dun, E. 1915. In: Gamble (Eds.) *Flora of the Presidency of Madras*. 1: 56.

Vajravelu, S. and Daniel, P. 1983. *Material for Carnatic Threatened Plants of India*.

Ramamurthy, V. 1983. In: Nair and Henry (Eds.). *Flora of Tamil Nadu*, Ser. I. Analysis 1: 29.

Ahmedulla, M. and Nayar, M. P. 1990. In: Nayar and Sastry (eds.). *Red Data Book of Indian Plants* 3: 87.

Nayar, M. P. and Sastry, A. R. K. 1990. Red data book of Indian plants Vol. III, Botanical Survey of India.

Doyle, J. J. and Doyle, J. L. 1987. A rapid DNA isolation procedure from small quantities of fresh leaf tissues. *Phytochem. Bull.* 19: 11-15.

Manickam, V. S. and Irudayaraj, V. 1988. *Cytology of ferns from Western Ghats South India.* Today & Tomorrows Printers and Publishers, New Delhi.

Bhojwani, S.S. and Bhatnagar, S.S. 1974. *The Embryology of Angiosperms*. pp. 107.

Gopalan, R. and Henry, A.N. 2000. *Endemic plants of India* Camp for the strict endemic of Agasthiamalai Hills, Southern Western Ghats, India.

Mohanan, N. and Sivadasan, M. 2002. *Flora of Agasthyamala.* Bishen Singh Mahendra Pal Singh Publication, Dehradun.

Brown, W. V. 1951. Apomixis in *Zepheranthus texana* Herb. *Amer. J. Bot.* 38(9): 697-702.

Craw, C. F. and Carman, J. G. 1987. Mechanism of apomixis in *Elymus rectisetus* from Eastern Australia and New Zealand. *Amer. J. Bot.* 74(4): 477-496.

Gupta, P. K. 1985. *Cytogenetics.* Rastogi Publications, Meerut, India.

Knox, R. B. and Heslop-Harrison, J. 1963. Experimental control of aposporous apomixis in a grass of the Andropogoneae. *Bot. Not.* 116: 127-141.

Stebbins G. L. 1950. *Variation and Evolution in Plants.* Columbia University Press, New York.

Richards, A. J. 1990. Studies in *Garcinia*, dioecious tropical forest trees: Agamospermy. *Bot. J. Linn Soc.* 103(4): 301-308.

Verheij, J. W. M. 1991. *Garcinia mangostana* L. In: Plant Resources of Southeast Asia. No. 2. Edible Fruits and Nuts, Eds. J.W.M. Verheij and R.E. Coronel, PROSEA, Bogor, Indonesia.

Bhat, V., Dwivedi, K. K., Khurana, J. P and Sopory, S. K. 2005. Apomixis: An enigma with potential applications. *Curr. Sci.* 89(11): 1879-1893.

Kaushal, P., Zadoo, S. N., Malaviya, D. R and Roy, A. K. 2005. Apomixis research in India: Past efforts and future strategies. *Curr. Sci.* 89(7): 1092-1096.

Hagerup, G. 1932. *Evolutionary Patterns and Processes in Ferns.* In: Proc. Advances in Botanical Research, pp. 330-336.

Advances in Pollen Spore Research Vol. XXVIII (2010) : 83-89
Editor : Dr. A. J. Solomon Raju
Today & Tomorrow's Printers and Publishers, New Delhi - 110 002 (India)

FLOWER-BUTTERFLY INTERACTIONS AND POLLINATION

G. Meerabai

Department of Botany,
Rayalaseema University, Kurnool 518 002,
Andhra Pradesh

ABSTRACT

The present study was carried out at Visakhapatnam located in Andhra Pradesh on the east coast of India. The plant species foraged by butterflies during the study period were identified and categorized into seven types on basis of the position of essential organs relative to the other floral organs, especially with anthers or pollen grains. Nectar sugars, amino-acid concentration, proteins in these flowers were studied by using the standard phyto-chemical methods. This study revealed that the tubular flowers are predominant with 'Sucrose' and containing high histidine-scale ranging from 4.00–6.00 representing the presence of more amino-acids. It is concluded that the tubular flowers avoid other insects except the long proboscoid butterflies, and provide strong nutritional resources to attract the attention of butterflies and accelerate chances for their breeding. The flower visits in all the plant species foraged by butterflies invariably end up in pollination.

Key words: Tubular flowers, sucrose, Histidine scale, pollination.

INTRODUCTION

In recent years there has been an increasing interest paid to the study of broad associations between floral features, physico-chemical characteristics of nectars and types of animal pollinators. Most of the nectar literature focuses on sugars, the major components of floral nectars. The amounts and sugar concentration of nectar have been related to pollinator type, especially considering the sucrose/hexose ratio. Amino acids in nectars and their evolutionary significance have also received attention. Baker and Baker 1973; 1982; 1986) proposed that the amount of amino acids present in nectars is directly connected with their pollination systems. These conclusions were disputed by Gottsberger *et al.* (1984; 1989). The debate has never been concluded on the basis of hard evidence from a large scale study. Today the "consensus view" is that plants adapted to pollination by butterflies have high concentration of amino acids compared to plants that are bird or bee pollinated (Gardner and Gilleman 2002). They emphasized the concentrations of amino acids to the taste of nectar. Hence, the adaptive significance of nectar components in relation to pollination by butterflies has now been studied and discussed.

METHODS OF STUDY

The present study was carried out at Visakhapatnam (Latitude 17° 42' and Longitude 82° 18' E) located in Andhra Pradesh on the east coast of India. The vegetation is mainly of the deciduous scrub jungle type dotted here and there with some evergreen species. In the rainy season there appears to be a luxuriant growth of a variety of herbs and shrubs which give a somewhat compactness to the vegetation.

During the study period, all the plant species visited by butterflies were collected and identified by a comparison with the authentic herbarium specimens available in the Department of Botany, Andhra University. The onset, progress, termination and the duration of blooming of each plant species were registered. Based on the flower structure, the flowers were categorized as zygomorphic, open and bowl shaped and tubular flowers. Taking into consideration the position of essential organs relative to the other floral parts, these were subdivided as (A) Flowers zygomorphic with the essential organs placed adjacent to or lying on the lower corolla lip, (B) Flowers zygomorphic with the essential organs oriented toward the upper lip, (C) Flowers open with the essential organs centrally positioned, (D) Flowers open with exposed numerous stamens, (E) Flowers tubular with the essential organs inserted, (F) Flowers tubular with the essential organs exserted and (G) Flowers with the essential organs rather elongated and oriented horizontally. The nectar accumulated in the flowers covered with butter paper bags was measured with graduated micropipettes at 10:00 and 17:00 h. In the case of flower which dropped off or closed by the afternoon, measurements were made at convenient hours. Sugar concentration was determined by using Refractometer. Analysis of nectar for the type of sugars present was done by Paper Chromatography (Harborne 1973). The presence and the relative amounts of amino acids in nectars were determined following the method of Baker and Baker (1973). Protein presence was indicated by the development of greenish blue colour when Bromo-phenol blue stain was placed on the nectar spot put and dried on the filter paper. The nectar feeding activity of butterflies was examined in relation to their role in pollination.

RESULTS

Table 1 gives the flowers type, nectar volume in micro litres (ìl), nectar sugar concentration (%), predominant sugar, amino acid score and protein presence for some plant species visited by butterflies. The quantity of nectar was found to be in traces in some and in some in measurable quantities. Nectar sugar concentrations varied from 12% to 62%. In 16 flower species 'sucrose' was found to be the dominant sugar, in another 28 species 'glucose' was so. All the plant species showed the presence of amino acids and the histidine score ranged from 0.0 to 6.0. The presence of proteins was recorded in all the plant species foraged by butterflies. The butterflies with varying lengths of proboscides, sizes and shapes invariably pollinated the open type and tubular and even cup-shaped flowers with nectar.

Table 1. Nectar volumes, concentrations at 10:00 and 17:00 hours, predominant sugar, Histidine scale and presence of proteins of different flower types

Name of the plant		Flower type	Nectar volume nant	Nectar concentra-	Predomi- nant	Histidine sugar		Proteins scale
		1000	1700	1000	1700			
Cadaba fruticosa	g	2.8	1.6	18	20	gSf	5.5	+
Capparis spinosa	d	2.0	1.02	26	21	gSf	6.0	+
Cleome viscosa	a	Traces	—	Gsf	0.0	+		
Sida cordifolia	c	Traces	—	Gsf	0.0	+		
Sida acuta	c	Traces	—	Gsf	0.0	+		
Tribulus terrestris	c	0.3	0.9	61	56	Gsf	2.0	+
Zizyphus mauritiana	c	0.02	0.02	57	51	Gsf	0.0	+

Z. oenoplia	c	0.02	0.02	53	49	Gsf	0.0	+
Sapindus emarginatus	f	1.3	1.6	39	26	Gsf	4.0	+
Anacardium occidentale	e	0.6	1.3	31	26	Gsf	3.0	+
Caesalpinia coriaria	f	0.02	0.06	48	26	Gfs	3.0	+
C. pulcherrima	g	0.68	1.22	29	18	gSf	6.0	+
Albizia lebbec	d	0.4	0.7	19	17	Gsf	0.3	+
Randia brandisii	f	0.3	0.4	31	22	Gsf	5.0	+
Ixora arborea	e	2.0	1.6	26	14	Gsf	4.0	+
Hamelia patens	c	0.9	1.3	26	21	Gsf	5.0	+
Murraya koenigii	e	0.06	0.01	23	16	Gsf	3.0	+
Borreria hispida	f	0.02	0.02	28	32	Gsf	4.0	+
Eupatorium triplinerve	f	0.8	0.4	28	23	Gsf	6	+
Tithonia rotundifolia	e	0.8	0.6	34	31	Gfs	1.0	+
Gaillardia picta	f	1.0	1.0	38	32	Gsf	5.0	+
Tridax procumbens	f	0.5	0.9	29	25	Gsf	5.0	+
Justicia procumbens	e	0.01	0.04	21	18	Gsf	1.0	+
Asystasia gangetica	e	0.02	0.2	22	18	gSf	0.0	+
Adhatoda vasica	b	3.8	3.9	21.5	18	gSf	1.0	+
Clerodendrum infortunatum	g	0.99	2.25	12	20	gSf	6.0	+
C. phlomidis	g	0.4	0.9	30	26	gSf	4.0	+
Lantana camara	e	0.06	0.04	18	23	gSf	4.0	+
Stachytarpheta indica	e	0.9	1.0	27	28	gSf	4.0	+
Premna latifolia	e	0.01	0.06	36	33	Gsf	2.0	+
Duranta repens	e	0.6	1.0	24	19	gSf	4.0	+
Citheroxylon subserratum	e	0.08	0.2	24	18	gSf	4.0	+
Tectona grandis	e	0.8	0.8	23	19	Gsf	3.0	+
Ocimum basilicum	b	0.01	0.01	18	16	Gfs	0.0	+
Hyptis suaveolens	b	0.02	0.04	24	19	Gsf	1.0	+
Jasminum angustifolium	e&f	0.5	0.61	28	16	gSf	4.0	+

Moringa oleifera	a	0.8	1.3	28	10	Gsf	1.0	+
Santalum album	c	Traces	—	Gsf	0.0	+		
Catharanthus roseus	e	1.0	2.2	25	22	gSf	6.0	+
Antigonon leptopus	c	0.02	0.4	58	54	Gsf	0.0	+
Jatropha gossypiifolia	c	1.9	0.2	62	59	Gsf	5.0	+
J. podagrica	c	1.4	1.8	56	53	Gsf	4.0	+

DISCUSSION

The amount of reward presented to a flower visitor is an important factor for the breeding system of the plants. The volume of nectar provided by a flower is just a secondary consideration. Primary preference is the concentration of energy giving chemicals and quality and quantity of amino acids in nectar (Neff *et al.* 1977; Baker 1978). Baker and Baker (1973) reported that butterfly nectars are not viscous. Pivnick and McNiel (1985) that *Thymelicus lineda,* a Hesperidae butterfly feeding on concentrated nectars (40–65 % sucrose) from a variety of flower species. This finding led them to question that butterflies feed primarily and most effectively on dilute nectars. The nectars of flowers now examined broadly fall into two ranges: 12–48% and 53–62%. The former being more common with frequent butterfly visited flowers whereas the latter is visited less often by butterflies. Probably as suggested by Pivnick and McNiel (1985) severe water stress may shift preferences to higher and lower concentrations respectively.

Baker and Baker (1973) and Cruden and Hermann-Parker (1979) demonstrated that butterfly nectars are sucrose rich. The present study showed that though most butterfly nectars are sucrose rich, they also utilized glucose rich nectars. The nectars of most tubular flowers contained more of the disaccharide, while those of open flowers its byproducts. These observations agree with Gottsberger *et al.* (1984), Percival (1961), Von Handen *et al.* (1972), Heinrich (1975) and Kapyla (1978). Nectar is a potential source of amino acids for the nutrition of butterflies. Naturally butterfly nectars are rich in amino acids. The mean of histidine scale score is 6.68 (Baker and Baker 1973 and Baker 1978). But, the present study showed that the nectars with low histidine scale are also visited by some butterflies. It reveals that butterflies need not necessarily feed on amino acid rich nectars.

The study revealed that the amino acid rich or poor and sucrose or glucose rich nectars are foraged by the butterflies. However, it is evident from the table that the tubular flowers are predominant with 'sucrose' and containing high histidine scale ranging from 4.00 to 6.00 representing the presence of more amino acids. This observation concludes that the flowers which are especially adopted for the visits of long proboscis butterflies with tubular corolla are providing strong nutritional resources for frequent visits to flowers and increase the chances for their pollination. This supported the view of Baker and Baker (1973) and Baker (1978). Based upon the phylogeny of the plant species examined, Baker and Baker (1973) suggested that amino acids have been constituents of nectars since the earliest stages of angiosperm evolution. It means that the composition of amino acids in nectars is shaped by phylogenetic constraints. Hence, the importance of plant phylogeny in nectar studies is becoming more apparent and the geographical focus is becoming wider. The nectar feeding activity of butterflies in all the examined plant species transfer pollen from flower to flower within and between individuals in order to contribute to pollination.

ACKNOWLEDGEMENT

I wish to record here my extreme indebtedness to late Prof. C. Subba Reddi, Department of Environmental Sciences, Andhra University, Waltair, for his invaluable and patient guidance as well as indispensable encouragement.

REFERENCES

Baker, H.G. and Baker, I. 1973. Some anthecological aspects of the evolution of nectar producing flowers, particularly amino acid production in nectar. In: V. H. Heywood (ed.) Taxonomy and Ecology, pp. 243-264, Academic press, London, NewYork.

Baker, H.G. and Baker, I. 1973. Amino acids in nectar and their evolutionary significance. Nature (Lond.) 241: 543-545.

Baker, H. G. 1978. Chemical aspects of the pollination biology of woody plants in the tropics. In: P.B. Tomlinson and M. H. Zimmerman (eds.), *Tropical trees as living systems*, pp. 57–82, Cambridge University Press, Cambridge.

Baker, H. G. and Baker, I. 1982. Chemical constituents of nectar in relation to pollination Mechanisms and phylogeny. In: M. H. Nitecki (ed.), *Biochemical aspects of evolutionary biology,* pp. 131-171, Chicago University of Chicago Press.

Baker, H. G. and Baker, I. 1986. The occurrence and significance of amino acids in floral nectar. Plant Systematics and Evolution 151: 175-186.

Cruden, R.W. and Hermann Parker, S. N. 1979. Butterfly pollination of *Caesalpinia pulcherrima*, with observations on a psychophilous syndrome. J. Ecol. 67: 155–158.

Gardner, M.C. and Gillman, M. P. 2002. The taste of nectar – a neglected area of pollination ecology. Oikos 98: 552– 557.

Gottsberger et al. 1984. Amino acids and sugars in nectar and their putative evolutionary significance. Pl. Syst. Evol. 145: 55 – 77.

Gottsberger et al. 1989. Variation in floral nectar amino acid with aging of flowers, pollen contamination and flower damage. Israel Journal of Botany 39: 167 – 176.

Harborne, J. B. 1973. *Phytochemical Methods.* Chapman and Hall, London.

Kapyla, M. 1978. Bionomics of five food nesting solitary species of bees (Hym. Megachillidae), with emphasis on flower relationships. Biol. Res. Res. Univ. Jyvaskyala 5: 3 – 89.

Neff et al. 1977. Flowers–Flower visitor system. In: G. H. Oriens and O. T. Solbrig (eds.), *Convergent evolution in warm deserts.* Stroudsburg, Dowden, Hutchinson and Ross, Inc.

Percival, M. 1961. Types of nectar in angiosperms. New Phytol. 60: 235 – 281.

Pivnick, K.A. and McNiel, J. N. 1985. Effects of nectar concentration on butterfly feeding: measurede feeding rates for *Thymelicus lineola* (Lepidoptera: Hesperidae) and a general feeding model for the adult Lepidoptera. Oecologia (Berl.) 66: 226-237.

Von Handen et al. 1972. The sugars of some Florida nectars. Am. J. Bot. 59: 1030-1032.

Advances in Pollen Spore Research Vol. XXVIII (2010) : 90-98
Editor : Dr. A. J. Solomon Raju
Today & Tomorrow's Printers and Publishers, New Delhi - 110 002 (India)

FLOWERING PHENOLOGY AND POLLINATION MECHANISM OF SOME MANGROVE PLANTS

K. Henry Jonathan and P. Vara Lakshmi

Department of Environmental Sciences,
Andhra University, Visakhapatnam 530 003

ABSTRACT

Mangroves are salt tolerant plants that typically grow in inter- tidal zones and are referred to as forests by the sea. Mangrove forests of Andhra Pradesh occur in Godavari and Krishna estuary inhabiting 35 mangrove species. The diversity of species in both the mangrove forests is different due to the change in geography of the location and disturbance from human intervention. Nineteen mangrove plants were studied to know the breeding mechanism, floral phenology and reproductive phenomena of different groups of mangrove flora. Majority of plants commence flowering in summer and post-monsoon periods and initiate fruiting subsequently. Mixed breeding system is the most successful reproductive mode adopted by a good number of mangrove species. The pollen carriers included bees, flies, butterflies, wasps, moths, birds.

Key words: Pollen vectors, cross-pollination, breeding systems.

INTRODUCTION

Mangroves are the plant communities occurring in the inter-tidal zones along the coasts of tropical and subtropical countries (Ramasubramanian and Ravishankar 2004). They are one of the most productive ecosystems and are typically an assemblage of flowering plants growing along the interface of marine and freshwater habitats (Selvam and Karunagaran 2004). They represent rich and diverse living resources and render livelihood and economy to fishermen communities and also help in the protection of coastal environment.

Globally mangroves are distributed across the tropical and sub-tropical forests and are predominantly found in the tropical region (Ramasubramanian et al. 2003). Asia and Australia have the greatest diversity and distribution of mangrove forests, more than 40 percent are found along the Asian coasts. Largest mangrove formations are found in Indonesia, Brazil and Sundarbans of India and Bangladesh (Ramasubramanian et al. 2003).

The total mangrove forest cover in India is 7 percent of the world's mangrove forest (MoEF Status Report 1987), whereas the total span of the world's mangrove forest covers 30 countries spreading about 99,300 sq km. The extent of the mangrove forest in India including the permanently under water is about 6,740 sq km (Status Report 1987). The mangrove forests in India occur predominantly in East Coast, West Coast and the Bay Islands with East coast contributing 56 percent of the total mangrove cover in India. The East coast mangrove forests are distributed in West Bengal, Orissa, Andhra Pradesh, Tamil Nadu and Andaman and Nicobar Islands (FSI Data) and are rich in flora and fauna diversity; while the west coast mangrove forests are present in the States of Maharashtra, Gujarat, Goa, and Karnataka (MoEF Status Report 1987).

The mangrove forests in Andhra Pradesh contribute 0.9 percent of the total forest area of the state (Ramasubramanian et al. 2003). The total area of the Godavari mangrove forest is 188 sq. km and very dense mangroves constitute 23 Sq km (FSI, 2003). In this context a brief report on some of the mangrove species were observed in the Coringa Mangroves of Andhra Pradesh and the same is discussed in view of the earlier studies reported on flowering phenology, reproductive biology and conservation by various authors. A comparison of the reproductive modes of a few species is examined depending on their mode of seedling dispersal and subsequent establishment in the mangrove forest.

MATERIALS AND METHODS

The mangrove forests in Andhra Pradesh are located in the estuaries of Godavari and Krishna River. The Godavari mangroves are located in the Godavari estuary of East Godavari district and lies between 16°30'–17°00'N and 82°10'-80°23'E. The present studywas carried out in Godavari and Krishna mangrove forests of Andhra Pradesh. Mangrove plants belong to several families but possess marked similarity in their physiological characteristics and structural adaptations. Godavari and Krishna mangroves have a record of 35 mangrove flora belonging to 27 genera of 20 families (Ramasubramanian et al. 2003) while 24 species belonging to 19 genera and 15 families were reported according to Rajasekhar et al. (2002). Godavari mangrove forest is spread over nine reserve forests that have varying densities and distribution of mangrove flora. All the nine reserve forests were surveyed for understanding the floral phenology and fruiting pattern of the mangrove flora. Observations were made for two successive years from 2005 till 2008 to record the flowering pattern and principal pollen vectors of the different mangrove flora in both the study areas. Frequent visits were made during the entire blooming period to record the insect vectors. The flower visitors were caught using nets and immobilized with ethyl alcohol and later identified. Some of them were washed in ethanol for the presence of pollen on their body parts to ascertain their role in pollination or otherwise pollen carriers. For the present study, 7 mangrove species were selected which included *Ceriops decandra* (Griff.) Ding Hou and *C. tagal* (Perr.) C.B. Rob. (Rhizophoraceae), *Aegialitis rotundifolia* Roxb. (Plumbaginaceae), *Aegiceras corniculatum* (L.) Blanco (Myrsinaceae), *Sonneratia apetala* Buch.-Ham. (Sonneratiaceae), *Derris trifoliata* Lour. (Fabaceae) and *Excoecaria agallocha* L. (Euphorbiaceae) from both the mangrove forests to understand and gather information on reproductive details such as breeding mechanism, flower visitors and their role as pollen vectors, analysis of floral rewards (pollen and nectar), floral phenology, seed set ratio and seedling dispersal mechanism. Biochemical aspects of flower nectar such as nectar sugar types, amino acids and protein estimation is carried as per Dafni et al. (2005). Pollen protein was quantified according to Lowry et al. (1951) to know the quantification of total protein content in pollen per flower.

RESULTS AND DISCUSSION

Of the selected plant species studied, *C. tagal* and *A. rotundifolia* occur only in Krishna mangrove forest and hence were studied in this forest. The other five plant species occur in Krishna (Latitude 15°42'–15°55'N and Longitude 80°42'– 81°01'E) and Godavari (Latitude 16°30'–17°00'N and Longitude 82°10'-80°23'E) mangrove forests but the study was conducted on these species in Godavari mangroves only. Of the 7 species examined, *Ceriops* species are true evergreen viviparous tree species, *A. rotundifolia* and *A. corniculatum* are evergreen shrubby crypto-viviparous tree species and the other three are non-viviparous species (Table 3). *S. apetala* is an evergreen tree species, *E. agallocha*, a semi-deciduous tree species while *D. trifoliata* is a deciduous woody climber. The last two species are distributed in oligo-haline to poly-haline zones. *A. corniculatum* is a meso-haline species, *C. decandra* and *S. apetala* poly-haline species while *C. tagal* and *A. rotundifolia* are eu-haline species.

Table 1. Nectar Sugar types present in the studied mangrove plant species.

Sugar type	*Ceriops decandra*	*Ceriops tagal*	*Aegialitis rotundifolia*	*Aegiceras corniculatum*	*Sonneratia apetala*	*Derris trifoliata*
Sucrose	+++	—	—	—	+++	+++
Fructose	++	+++	+++	+++	++	+
Glucose	++	++	+++	++	++	++

Table 2. Nectar amino acids of the studied mangrove plant species

S. No.	Amino acid type	*Ceriops decandra*	*Ceriops tagal*	*Aegialitis rotundifolia*	*Aegiceras corniculatum*	*Sonneratia apetala*
1	Alanine	++	+++	+++	+++	++
2	Arginine*	+	—	++	+++	
3	Aspartic acid	++	++	++	++	+++
4	Cysteine	+++	+++	++	+++	++
5	Cystine	++	+++	+++	+++	+++
6	Glutamic acid	+	+	++	+++	+++
7	Glycine	++	+++	+++	+++	+++
8	Histidine*	+	—	++	+++	+
9	Iso-leucine*	—	—	—	—	—
10	Leucine*	—	—	—	—	—
11	Lysine*	—	+	++	+++	++
12	Methionine*	—	—	—	—	—
13	Phenly-alanine*	+	—	+	—	—
14	Proline	—	—	+++	++	+++
15	Serine	++	+++	+++	+++	+++
16	Threonine*	+	++	+++	+++	—
17	Tryptophan*	+	—	++	—	—
18	Tyrosine	++	++	+	++	+++
19	Valine*	++	—	+++	—	—

+++: dominant, ++: moderate, +: traces, —: absent. *essential amino acids for insects.

Table 3: Mode of reproduction of selected mangrove species

Viviparous	Crypto-viviparous	Non-viviparous
Ceriops decandra	*Aegiceras corniculatum*	*Sonneratia apetala*
**C. tagal*	**Aegialitis rotundifolia*	*Derris trifoliata*
		Excoecaria agallocha

* Species present in Krishna mangrove forest, not confirmed in Coringa forest.

All the species, except *E. agallocha* are hermaphroditic, self-compatible and exhibit mixed breeding system. *E. agallocha* is dioecious and obligately out-crossing. In all, the flowers offer nectar and pollen as rewards to flower visitors. The nectar volume, sugar concentration and sugar types were examined for all the species except *E. agallocha*; and nectar amino acids for five species except *D. trifoliata* and *E. agallocha*. Nectar is sucrose rich in *C. decandra*, *S. apetala* and *E. agallocha*; fructose rich in the following species *C. tagal*, *A. rotundifolia* and *A. corniculatum* while the *A. rotundifolia* is also rich in glucose (Table 1). These nectar characteristics are suited for entomophily. Insects require ten essential amino acids and some of them are found in these plant species - arginine, histidine, phenylalanine, threonine, tryptophan and valine in *C. decandra*; lysine and threonine in *C. tagal*; arginine, lysine, phenylalanine, threonine, tryptophan, valine and histidine in *A. rotundifolia*; arginine, lysine, threonine and histidine in *A. corniculatum*; and arginine, lysine and histidine in *S. apetala* (Table 2). Further, pollen also contains protein, the amount of which varied with each plant species. The floral rewards are therefore important nutrient sources for different species of insect pollinators. Quantitative estimates of floral rewards of the mangrove plant species is given in (Table 4).

Table 4. Quantitative estimates of floral rewards of the mangrove plant species.

Plant species	Pollen protein content/flower (μg)	Nectar protein content/flower (μg)	Nectar sugar content/flower (mg)	Pollen ovule ratio	Pollen size (μm)
C. decandra	24.0	—	0.66±0.1	2135:1	16.6
C. tagal	21.6	48.11±10.52	2.79±0.5	2446:1	15
A. rotundifolia	30.04	—	3.60±0.49	1443:1	119.52±10.49
A. corniculatum	28.0	—	1.73± 0.39	2174:1	33.2
S. apetala	1.71 mg	13.94±2.20	1.99±0.51	420:1	41.5
D. trifoliata	6.0	—	0.43	1873:1	41.5
E. agallocha	9.0	—	—	35,259.4:1	33.2

In *C. tagal*, the flowers are specialized and possess explosive pollination mechanism which is adapted for pollination by insects and also by wind while in *C. decandra*, the flowers are not specialized, have simple floral mechanism and are adapted for pollination by bees and wasps. In *A. rotundifolia* and *A. corniculatum* also, the flowers are not specialized and promiscuous for foraging by any insects; the former species is characteristically melittophilous while the latter species is entomophilous, and in both the species, anemophily also occurs to some extent. The floral characteristics of *S. apetala* suggest chiropterophily but morning anthesis and the presentation of floral rewards during daytime facilitate foraging by day-active foragers only and accordingly, the plant is found to be primarily melittophilous and partly anemophilous. *D. trifoliata* flowers are specialized and exhibit explosive pollination mechanism which is adapted for tripping and subsequent pollination by insects. The short- and long-tongued bees trip the flowers and hence the plant is melittophilous. *E. agallocha* with male and female flowers on separate individuals is found to be both anemophilous and entomophilous. In sampled sites, the male trees are double the number of female trees, the ratio of male to female flowers is 16:1 and each male flower produces copious amount of pollen. These characteristics are adaptations for anemophily. The long catkins of male trees and the short mixed cymes of female trees with a number of flowers and the presence of several such inflorescences on each tree simultaneously are quite attractive to foragers. The yellow stamens and bright green shining styles appear to enhance attractiveness to foragers. The nectar is in traces in male flowers while it is relatively measurable in female flowers. These characteristics indicate that the plant is also evolved for pollination by insects. Insect-pollination is considered to be more economical than anemophily; in the latter mode of pollination, there is wastage of huge amount of pollen and also it is expensive for the plant energetically.

In all the studied species except *E. agallocha*, autogamy is a mode of pollination but fruit set rate in this mode is negligible in *Ceriops* species and *D. trifoliata*, and more in other species, but it is highest in *S. apetala*. In *E. agallocha,* anemophily assures the success of sexual reproduction in the absence of insect pollinators. Therefore, the ability to out-cross to set fruit through anemophily in *E. agallocha* and the ability to self-pollinate and set fruit in the absence of insect pollinators in all the other plant species seems to be a "fail-safe" strategy for reproductive assurance during colonism or when conspecifics do not occur nearby. Mangrove areas are known for their harsh and uncertain environmental conditions and the ability of the studied plant species to reproduce sexually in the total absence of insect pollinators is highly adaptive for their survival and colonization.

Flower bud abortion occurs in *C. decandra, A. corniculatum, S. apetala* and *D. trifoliata*; it is highest in the first species. Natural fruit set rate is 16% in both the species of *Ceriops*, 55-61% in *A. rotundifolia, A. corniculatum* and *S. apetala*, 29% in *D. trifoliata* and 92% in *E. agallocha*. In the last species, fruit predation is 25%; a beetle, *Chrysocoris patricius* consumed the fruits prior to their fall from the mother plant. Single-seeded fruit is the characteristic in *Ceriops* species in which each ovary contains 6 ovules and in *A. corniculatum* in which each ovary contains 35 ovules. *A. rotundifolia* flower is single-ovuled and produces a single seed. In *S. apetala*, each fruit produces an average number of 118 seeds against an average number of 488 ovules per ovary and fruit abortion takes place to an extent of 11%. In *D. trifoliata*, 1-3 seeds are produced per pod although the ovary contains 6 ovules. The recorded natural fruit set rates, ovule abortion, fruit abortion and reduced seed set rate were considered to be related to the pollination rate, elimination of genetically inferior seeds prior to their maturation and to the availability of nutrient resources to the parent plant.

In all the studied species, seeds are not dormant. In *Ceriops* species, the seed produces naked seedling while it is still attached to the parent plant; the seedling is small and erect in *C. decandra* but it is longer and in hanging position in *C. tagal*. In both the species, a yellow collar is visible separating fruit from the seedling. In *A. rotundifolia* and *A. corniculatum* also, the seed produces seedling within the fruit pericarp while it is still attached to the parent plant; the seedling is erect in the former and hangs downwards in the latter species. The seedling of *A. rotundifolia* is relatively larger than that of *A. corniculatum*. These are important field characteristics to identify the plants during their fruiting season. The fruits hang downward in *S. apetala*, are horizontal or vertical in *D. trifoliata*, and erect in *E. agallocha*. In these non-viviparous species, seed is the propagule and it does not germinate on the parent plant but germinates when settled in the parental or other sites. In all the species except *C. tagal*, seedling or seed dispersal and establishment occurs through self-planting and stranding strategies. In *C. tagal*, seedling dispersal and establishment occurs through self-planting strategy only. The self-planting strategy is effective in undisturbed sites while the stranding strategy is more effective in exploited and open areas of mangroves.

The study suggests that the characteristics such as sexual system, mixed breeding system, self-pollination with or without wind or insects, vivipary, crypto-vivipary and non-vivipary, self-planting and stranding strategies are adaptations in the studied plant species for survival and colonization in mangrove areas. The plants are important nectar and pollen sources for different insect species for their own survival and for raising their offspring. The relationships between mangrove plants and insects are mutualistic, the former for reproductive success and the latter for food. Therefore, these relationships are important for structuring and sustaining mangrove forests.

REFERENCES

Dafni, A., Kevan, P.G. and Husband, B.C. 2005. *Practical Pollination Biology*. Enviroquest Ltd., Ontario.

Lowry, O.H., Rosebrough, N.J., Farr, A.L. and Randall, R.J. 1951. Protein measurement with the folin phenol reagent. *J. Biol. Chem.* 193: 265-275.

MOEF Status Report – Mangroves of India, 1987.

Rajasekhar, P.S., Brahmaji Rao, P. and Subba Rao, M.V. 2002. Biodiversity values and traditional utilization pattern of Godavari mangroves, Andhra Pradesh. *Proc. National Seminar on Conservation of Eastern Ghats.* pp. 232-238, EPTRI, Hyderabad.

Ramasubramaniam, R., Ravishankar, R. and Sridhar, D. 2003. Mangroves of Andhra Pradesh: Identification and Conservation Manual, M.S. Swaminathan Research Foundation, Chennai.

Ramasubramanian R. and T. Ravishankar 2004. Mangrove forest restoration in Andhra Pradesh. M.S. Swaminathan Research Foundation, Chennai.

Selvam V. and Karunagaran, V.M. 2004. Ecology and biology of mangroves-coastal wetlands: mangrove conservation and management, M. S. Swaminathan Research foundation, Chennai.

Advances in Pollen Spore Research Vol. XXVIII (2010) : 99-129
Editor : Dr. A. J. Solomon Raju
Today & Tomorrow's Printers and Publishers, New Delhi - 110 002 (India)

FUNGAL AIRSPORA OF VISAKHAPATNAM

Saraswathi Rao

Department of Botany,
Dr. B. R. Ambedkar Open University
Prof. G. Ram Reddy Marg, Road No. 46, Jubilee Hills
Hyderabad 500 033

INTRODUCTION

The fungi are highly diverse and versatile organisms adapted to all kinds of environments. They form an important component of the decomposer unit in the ecosystem, either terrestrial or aquatic. They produce large numbers of spores which serve, in the majority of cases, as a means of dispersal mainly through the atmospheric route. This spore dispersal is necessary for the fungus concerned to maintain its population within its existing range and to colonize new habitats. Besides, the dispersal of spores acts as "an air mail service transmitting genes between established mycelia" (Gregory 1952a). The result of the dispersal process through the medium of atmosphere is that, in practically all situations, the air contains many kinds and large numbers of spores. To describe the population of airborne fungal spores and other biogenic particles, Gregory (1952b) introduced the useful concept of the 'air spora' an inescapable phenomenon of nature. Therefore, airspora deserves a study in its own right as a natural phenomenon (Gregory 1973). Its study is of much relevance to the understanding of the ecology of fungi (Ingold 1971). Furthermore, there has been an increasing interest in the study of airspora because of its applications in medicine (inhalant allergy), agriculture (epidemiology of air borne plant diseases), bio-deterioration and spoilage of food stuffs and other useful materials, and air pollution (Gregory 1973; Edmonds 1979; Shivpuri 1982).

Keeping in view the above mentioned significance of the study of airspora and its applications, the fungal airspora of Visakhapatnam, a fast developing industrialized city, was earlier investigated by Subba Reddi (1970) and, Atluri and Subba Reddi (1981). These workers used a wind impaction sampler – a 0.53 cm diameter glass rod, wrapped with an adhesive coated 18 mm square cellophane strip and exposed vertically, for trapping the air spora. Although such a sampler is useful for trapping a variety of particles in the outdoor atmosphere, its basic disadvantage is its change in efficiency with wind speed and particle size (Gregory 1973; Ogden et al. 1974). Its efficiency is very much less with respect to particles of small size which escape collection, leaving larges types to dominate the catch. Also, conversion of counts to concentration is a tedious process and requires additional data on wind speed. As identification is by visual means based on spore characteristics, species delineation is nearly impossible except with certain plant pathogens. In these works, as was also the case with most of the surveys of the aero-mycoflora followed by visual identification technique, the airborne conidia of aspergilla were placed together with penicillia in a form group 'aspergilli' because species but not genera are involved in biological activity, the occurrence of species of these fungi in the air is indefinite need of more study (Kramer et al. 1960; Vittal an Ponnusamy 1982). Such studies may become increasingly important in clinical allergy since antigenicity may differ from species to species (Austiwick 1966; Agarwal et al. 1968).

Therefore, in order to supplement and update the information on the fungal airpsora of Visakhapatnam, already made available by using wind impaction sampler (Subba REddi, 1970; Atluri and Subba Reddi, 1981), the present cultural study was undertaken using the Andersen sampler (Andersen, 1958), which is known for its value in the volumetric collection of viable airborne microorganisms, particularly those with small spores like *Aspergillus, Penicillium* and *Cladosporium*, etc. as expected from the collection efficiency of the sampler, the small spored fungus aspergillus dominated in the present study and contributed ca. 43 % to the total spore catch. Among the 28 species identified, *A. luchuensis* dominated and contributed ca. 59 % to the total genus isolates; the second predominated species was *A. violaceo-fuscus* (ca. 11 %). *Cladosporuim*, which emerged as the dominant component of the fungal airspora of Visakhapatnam in earlier studies with a contribution of ca. 29 % to the total spore catch (Atluri and Subba Reddi 1981), is now relegated to the second position with a contribution of ca. 25 %. Among the six species identified, *C. herbarum* dominated with a contribution of 70 % to the total genus isolates. *Pencillium*, which is known to dominate *Aspergillus* in the temperate environments (see Rati and Ramalingam 1976), occupied third position and contributed ca. 10 % to the total fungi isolates. These and other data are presented in this paper. The same are discussed in the light of comparable studies from India including the earlier study from Visakhapatnam and abroad.

MATERIALS AND METHODS

The present study was carried out at Visakhapatnam from November 1984 through October 1986. Sampling was done on the top of Erskine College building 9.14 m high on the Andhra university campus. This building is centrally located on the campus and is approximately 2 km from the sea shore. The campus a relatively elevated region has huge buildings. Towards south west there is a botanical garden with large trees, thickets and other vegetation. On the northern side a large area of highly eroded laterite soil is present and is covered by sparse vegetation for about 3 km, while on the eastern side there is a vegetation cover for about 1 km followed by sea shore. The most dominant soil types in the region are lateritic red loams. They are coarse textured, highly permeable and well leached. They generally contain very low humus content and are very poor in fertility. Moreover they are easily subjected to water and wind erosion resulting in an irregular topography and numerous gullies are cut into the soil profile.

The Andersen sampler (Andersen 1958) was employed for sampling airborne microorganisms. It is one of the samplers recommended by aerobiology group of IBP (see Edmonds and Benninghoff 1979). It is a suction type impaction sampler through which air is drawn at the rate of 28.3 liters per minute. Air is sucked into the sampler through a circular orifice and through a series of six circular plates each having 400 perforations, through which air along with bio-particles, is impacted directly onto the sterile medium in petri dishes. Succeeding plates in the set have perforations of progressively diminishing size. The orifice diameter of stage 1 is 1.81 mm, that of stage II 0.91 mm, stage III 0.71 mm, stage IV 0.53 mm, stage V 0.34 mm, stage VI 0.25mm. the range of particle sizes that can be trapped at each of these stages respectively are 7.0, 4.7-7.0, 3.3-4.7, 2.1-3.3, 1.1-2.1 and 0.65-1.1 μm. The wall losses in this spore trap are negligible while retention is 100% (Andersen 1958). The orifice stages of the sampler are cleaned and disinfected with 90% isopropyl alcohol before and after the use of the instrument. After placing the petri dishes in order and holding them in place by metal spring fasteners, the vacuum pump was connected to the sampler. After sampling, the petri dishes were inverted and placed in covers. Conversion of counts to number per cubic meter of air was done by using the following formula:

$$\frac{\text{Total number of colonies in 6 plates}}{\text{Total volume of air in liters}} \times 100$$

The interrelationship between the airborne fungal populations and the meteorological factors was determined using the statistical techniques such as simple correlation, partial correlation, simple correlation coefficient was calculated to reflect the degree of relationship between the number of airborne fungi and each meteorological variable. Partial correlation coefficients were calculated to know the actual effect of each weather variable on fungal concentrations by suppressing the effect of other variables.

After a preliminary study as to the choice of the medium, Sabouraud's dextrose agar was chosen because it gave the maximum isolates. Penicillium at 20 units/ml and streptomycin at 40 μg/ml were added to the sterilized medium to suppress bacterial and actinomycete growth. Samples were collected at four different ecological sites viz., 1. Posh residential area-marked by well spaced out houses, with big gardens and with minimum of dirt, 2. Jagadamba centre - a busy shopping and entertainment centre. The main feature of this area is the presence of concrete buildings, with practically no vegetation, dense population and heavy vehicular traffic. 3. Purna market - the most important market for fruits, vegetation and cereals. This public place is crowded with stalls and people. Plenty of garbage is seen around. 4. Industrialized site-in the vicinity of Coromandel Fertilizers – this is 40 km from the city. Sampling during 'Polamamba Mela' was carried out. Air was sampled to study the effect of human congregation of air borne fungi. Air was sampled a day before mela, o during the festival and the day after the mela. The mela is celebrated to honor the presiding deity, Polamamba at the temple near Andhra university campus and is usually attended by a very large number of people.

The time of sampling used in the present study was chosen after knowing the temporal distribution of air borne fungi at 2-h intervals. The sampling time was around 1000h. The duration of sampling was 3 min and was selected after conducting a series of sampling tests to see the largest number of isolates without overcrowding. The Petri dishes with the medium after the exposure in the sampler were incubated at room temperature (20-35 C) for six days after which the final count of the isolates was taken. The plates were examined and the isolated colonies were characterized. Their cultural and microscopic characteristics were noted and used for their identification, which was done with the aid of standard descriptions of fungi given by Ellis (1971; 1976), Subramanian (1971), Raper and

Thom (1968) and Raper and Fennel (1973).

RESULTS

Forty eight air samples, collected over a period of two years, yielded 4314 fungal isolates. Pure cultures of these isolates were maintained, subsequently characterized and identified. The data were analyzed for species composition of the air borne fungal flora, relative abundance, seasonal variations, spatial variations, effects of weather factors and human activity. Among the fungi isolated, 84 species in 29 genera, 41 non-sporulating and 2 yeast types were recognized (Table 1). Over the 2 yr period, no isolate of basidiomycetes was encountered. Among the other three classes of fungi, deuteromycetes preponderated with an average share of 90.47%, followed by ascomycetes 2.2% and phycomycetes 1.46%. The various fungi identified in each of the two years of study, their annual totals and percentage contribution to the total fungal isolates, the two year average and the average percentage contributions are given in Table-1. *Aspergillus* was the predominant genus of the study. Isolates of this genus on average made up 42.88% of the total fungal isolates. The second dominant genus was *Cladosporium,* with a contribution of 25.06%. then followed in the descending order *Penicillium* 9.93%, the non-sporulating fungi 5.73%, *Curvularia* 4.58%, *Alternaria* 1.57%, *Drechslera* 1.41%, yeasts 1.39%, *Veronaea* 1.14% *Chaetomium* 0.80%, *Paecilomyces* 0.75%, *Mucor* 0.70%, *Nigrospora* 0.64%, *Periconia* 0.57%, *Rhizopus* and *Stachybotrys* 0.43% each, *Bispora* 0.27%, *Scopulariopsis* and *Trichothecium* 0.23% each, *Absidia* and *Memnoniella* 0.21% each, *Diplococcum* and *Myrothecium* 0.18% each, *Gliocladium* 0.14%, *Alysidium* 0.10%, *Bidenticula* 0.09%, *Syncephalastrum* 0.07%, *Fasarium, Phytophthora*, *Humicola, Verticillium* 0.02% each. In *Aspergillus* 28 species were recognized within in genus of which, *A. luchuensis* was the dominant one with a contribution of 58.93% to the total of the two years mean of the genus isolates. Then followed in the descending order are *A. violaceo-fuscus* 11.42%, *A. niger*, 5.63%, *A. nidulans*, 3.65%, *A. flavipes* 2.92%, *A. parasiticus* 2.74%, *A. flavus* 2.63%, *A. candidus* 2.32%, *A. niveus* 2.23%, *A. fumigatus* 1.63% , *A. repens* 1.27%. The other 17 species, each contributed less than 1%. *Cladosporium* genus was represented by six species. *C. herbarum* with a contribution of 70.17% of the total of 2 years mean of the genus isolates outnumbered others and was followed by *C. cladosporioides* 18.15%, *C. macrocarpum* 4.38%, *C. oxysporum* 2.71% and *C. chlorocephalum* 1.81%. The unidentified species comprises of 2.78% of the genus isolates. *Penicillium* represented 14 species which collectively comprised 94.05% of the mean total genus isolates. Of these species, *P. fellutanum* was relatively more common and made up of 29.01% and was followed by *P. rubrum* 11.22%, *P. atramentosum* 10.26%, *P. citrinum* 7.76%, *P. oxalicum* 7.57%, *P. variabile* 6.92%, *P. implicatum* 4.53%, *P. italicum* 4.11%, P. lanosum and *P. purpurogenum* 3.96% each, *P. steckii* 1.84%, *P. granulatum* 1.61% and *P. frequentas* and *P.viridicatum* 0.65% each. *Curvularia* genus represented 9 species of which *C. lunata* dominated and accounted for 42.69% of the genus isolates. Next in order were *C. indica* 18.88%, *C. maculans* 10.86%, *C. penniseti* 7.60%, *C. interseminata* 6.93%, *C. ovoidea* and *C. tritici* 5.52% each, *C. pallescens* 1.50% and *C. geniculata* 0.50%. *Alternaria* genus was represented by 7 species. *A. dianthicola* was relatively more abundant and accounted for 30.17% of the mean total of genus isolates. This was followed by *A. tenuissima* 24.33%, *A. citri* 18.98%, *A. papaveris* 10.22% and *A. alternata* 10.46%, *A. fasciculata* and *A. longipes* 2.92% each. *Drechslera* represented 5 species. *D. australiensis* was relatively dominant with a share of 29.08% in the mean total of genus isolates. This was followed by *D. poae* 27.45%, *D. tripogonis* 20.92% and *D. hawaiiensis* 9.78%. Two species each of *Nigrospora, Memnoniella, Paecilomyces* and *Stachybotrys* were identified. Within *Nigrospora, N. sphaerica* made up 85.63%, within *Memnoiella, M. levispora* contributed 54.55% within *Paecilomyces, P. varioti* comprised 96.94% and within *Stachybotry*s, *S. atra* accounted for 94.96% of the isolates. Excepting these 10 genera, all others identified were represented each by a single species. The average percentage contributions of these species therefore were the same as shown against their genera (Table 1).

An examination of the month wise distribution of total fungal colonies made it possible to divide a year, mycologically into two parts: October-March and April-September respectively, recording a higher and lower frequency of air borne viable fungi. A seasonal trend though not well defined was evident in the distribution of the more abundant genera of fungi. Although the same trend was not exactly repeated during the two years, the general pattern remained the same. Based on their distribution pattern, the fungi encountered could be classified under the following categories.

1) Those occurring round the year, *Aspergillus, Penicillium* yeasts, *Rhizopus* and non sporulating colonies come under this category.

2) Those occurring over a major part of the year. *Cladosporium, Curvularia, Drechslera, Alternaria* are the representatives.

3) Those occurring over a well defined period. *Gliocladium, Nigrospora, Periconia, Chaetomium, Veronaea, Bispora* represent this category.

4) Those occurring sporadically throughout the year. These included *Absidia, Mucor, Myrothecium, Memnoniella, Paecilomyces, Scopulariopsis,*

Stachybotrys.

5) Those occurring rarely. P*hytophthora, Syncephalastrum, Alysidium, Bidenticula, Diplococcium, Humicola, Fusarium, Stachybotrys, Verticillium* come under this group.

The genus *Aspergillus* with its year round occurrence was more frequent during January- February and also to some extent in May; its lowest record was in March. Nine of the 28 recognized species of the genus occurred more or less throughout the year. These species included *A. luchuensis*, *A. flavipes*, *A. violaceo-fuscus*, *A. parasiticus, A. flavus, A. nidulans, A. niger, A. fumigatus* and *A. niveus*. The predominant species *A. luchuensis* was encountered in all the samples collected and its distribution followed quite closely that exhibited by the genus. The other 8 species exhibited seasonal periodicities of their own, with peak incidence of *A. parasiticus* in January; *A. flavus* during October-November; *A. niger* during May-June; *A. nidulans* in March; *A. fumigatus* and *A. flavipes* in July; *A. violaceo-fuscus* in July and of *A. niveus* in June. The other 12 species did not display any seasonal pattern in their incidence in the air.

Penicillium with its year round occurrence exhibited three periods of higher incidence. They were March –April, July-August and October-November. The first being the period of highest incidence. The species which occurred more commonly and contributed to the peak were *P. implicatum, P. atramentosum*, *P. purpurogenum* and *P. italicum*. There was no distinct periodicity in the incidence of other species.

Yeasts with year round occurrence, frequented more in December. Non-sporulating fungi occurred in increased numbers from July to December and in March. Peaks were evident in November. *Rhizopus nigricans* was observed throughout the year, with a higher frequency in December.

Among the fungi trapped over a major part of the year, *Cladosporium*, the second dominant genus of the study, occurred over a 10 month period from July- April, with a greater frequency during November-February and with a peak in January. This seasonal trend was also evident in the occurrence of the dominant species *C. herbarum*. The other species *C. chlorocephalum, C. marcocarpum* and C. *oxysporum* also displayed more or less the same trend in their occurrence in the air. *C. cladosporiodes* was seen from December to March with a peak in February.

Alternaria was recovered from October-March and again from July-August, the month of November recording the highest concentration. The seasonality of *A. dianthicola, A. papeveris, A. citri* and *A. tenuissima* was quite similar to that of the genus. The peak concentration of *A. dianthicola* was recorded in August. *A. fasciculata* and *A. longipes* appeared for one or two months only. *Curvularia* was trapped over 11 months from May-March, with a greater frequency in October and again in January. The most prevalent species *C. lunata* quite closely followed the seasonal pattern of the genus, except that it was observed in peak concentrations in July-August period. Other species were seen to occur irregularly. *Drechslera* was trapped in 11 months of the year from June-April, with relatively higher concentration in November and in January. No definite trend was evident with individual species.

A number of genera occurred over a well defined period in the year. The genus *Nigrospora* was isolated from August-December, with a greater frequency in September. A quite similar trend was reflected in the incidence of *N. sphaerica*. The other species *N. oryzae* was recovered during September-October of 1984-85. *Gliocladium roseum* was seen from February-June with a relatively higher incidence in May. *Periconia jabalpurensis* was seen from March-April and from June- September, with greater frequency in July. *Chaetomium globosum* was seen from March to October (excepting April) with a relatively higher frequency in September. *Veronaea apiculata* frequented from August to December, with a peak in October. *Bispora* was seen from May-September and in February. Some fungi occurred sporadically throughout the year without any trend of seasonal periodicity. They were *Mucor hiemalis*, *Scopulariopsis brumptii* and *Stachybotrys atra* with higher frequency in September, *Paecilomyces varioti* in March and October, *Myrothecium roridum* in January and March, *Absidia heterospora* during June-July and *Trichothecium roseum* in October.

As already stated the fifth category of fungi included those fungi that occurred rarely. *Memnoiella levispora* occurred in March and April of 1984-85. Some fungi were seen only twice during the period of study. *Syncephalastrum racemosum* in March and in July of 1985-86, *Bidenticula cannae* in June 1984-85 and May of 1985-86. *Alysidium resinae* in August of 1984-85 and March of 1985-86, *Diplococcium spicatum* in March of 1984-85 and July of 1985-86, *Memnoniella echinata* in January and July of 1985-86. There were other fungi seen only once in the course of study. They included *Humicola brevis* in October of 1984-85, *Fusarium moniliforme* in January of 1984-85, *Verticillium terrestre* in March of 1985-86, *Paecilomyces fusisporus* in October of 1984-85. *Phytophthora* appeared in March 1984-85. When the distribution of fungi in air is considered in relation to the climatic seasons, it can be concluded that the cold season (mid November- mid February) recorded the highest percentage of the fungal isolate in both years of study, followed by the post monsoon season (mid September- mid November), the pre monsoon (hot weather) season (mid February-May), and south west monsoon (June- mid September).

Total fungi frequented more at the Purna market with 1680 colonies/m^3

of air. Then followed in the descending order were the industrial site (1275/m^3), posh residential area (1260/m^3) and Jagadamba centre (970/m^3). In all, eight categories of fungi were recognized from these samples. These included *Aspergillus, Cladsporium, Penicillium, Alternaria, Curvularia*, non-sporulating fungi, yeasts and the phycomycete fungi. All these were encountered at the four sites, except *Curvularia* which was absent at the Purna market site.

Aspergillus was the predominant genus at all the four sites. Its incidence was the highest at the Purna market (960/m^3). The concentrations at the industrial site (170/m^3) and Jagadamba centre (465/m^3) were approximately half, and those at posh residential area (345/m^3) were one third of the concentrations observed at Purna market site. The second most abundant genus at the industrial (390/m^3) and posh residential (245/m^3) sites was *Cladosporium* while *Penicillium* was so at the Purna market (290/m) and Jagadamba centre (145/m). The third important genus was *Curvularia* at the industrial site (155/m^3) that at the posh residential (160/m^3) was the non-sporulating fungi, while that at Jagadamba centre was *Cladosporium* (115/m^3), and that at Purna market (160/m^3) were the yeasts. The fungus that occupied the fourth place in the order of relative abundance at the industrial site was *Alternaria* (115/m^3) at the posh residential area it was *Curvularia* (95/m^3) at Jagadamba centre it was yeasts (60/m), while at the Purna market the Phycomycete fungi occupied this position with 110/m^3. The fifth important genus at the industrial and posh residential sites was *Penicillium* with 105/m^3 and 70/m^3 respectively. The non-sporulating fungi with 35/m^3 assumed this position at the Jagadamba centre, while at the Purna market it was *Cladosporium* (75/m). Mycelia sterilia occupied the sixth position at the industrial (70/m^3) and Purna market (25/m^3), but at the post residential site it was yeasts (65/m^3) and at the Jagadamba centre it was the *Phycomycete* fungi (30/m^3). Yeasts ranked seventh at the industrial site (50/m^3), *Curvularia* at Jagadamba centre (15/m^3), *Alternaria* at posh residential (50/m^3) as well as purna market sites (20/m^3). The lowest concentrations encountered at the industrial (10/m^3) and posh residential sites (32/m^3) were of Phycomycete fungi, while those at Jagadamba centre were of *Alternaria* (10/m^3).

Analysis for simple correlation between temperature and total fungal isolates gave a negative value of the coefficient which is rather significant for 1984-85 (r= -0.5544) and not significant for 1985-86 (r=-001975). The correlation coefficient for *Aspergillus* isolates was also negative, but not significant in both years (1984-85), (r=-0.2971); 1985-86 (r= -0.1622). Likewise the partial correlation coefficient for total fungal isolates was also negative, and significant for 1984-85 (r=-0.6702) and not significant for 1985-86 (r=-0.1460). So also was the relationship between *Aspergillus* and temperature (1984-85, r= -10.4545, r= -0.1273.

Simple correlation between RH and total fungal flora, and RH and *Aspergillus* isolates gave a positive relationship. But in the case of total flora, the value of coefficient was not significant for 1984-85 (r= 0.0359) while it was significant for 1985-86 (r= 0.1129). On the other hand, in the case of *Aspergillus* the coefficient values for both years of study were significant (1984-85, r= 0.1283); (1985-86, r= 0.2873). Partial correlation between RH and total fungal flora, and RH and *Aspergillus* isolates also showed a positive relationship in the case of total fungus flora. Unlike the values of simple correlation the coefficient value of 1984-85 was significant (r= 0.3313) whereas that of 1985-86 was not significant (r= 0.0226). In the case of *Aspergillus* the coefficient values obtained for both years (1984-85, r= 0.4478), 1985-86, r= 0.3112) were rather more significant than those of simple correlation coefficient.

Simple correlation of total fungal isolates and wind speed gave a positive significant value of coefficient for 1984-85 (r= 0.3777) but negative not significant for 1985-86 (r=-0.1916). Similar was the trend of relationship between *Aspergillus* isolates and wind speed (1984-85, r= 0.4813); (1985-86 r= 0.2423). The partial correlation coefficient in the case of total fungal flora was also positive and significant (r=0.5877) for 1984-85 but negative and not significant (r=-0.1489) for 1-85-86, but in the case of *Aspergillus* the values of both years were positive and significant (1984-85, r= 0.6460; 1985-86, r= 0.3437).

Three air samples were collected on each day, before, during and after Polamamba mela to study the effect of human activity on the air spora concentrations. As expected, the fungal isolates were more numerous (1110/m^3 of air) during mela than those encountered on the day before (680/m^3 of air) and on the day after the mela (700 /m^3 of air). The three air samples yielded *Aspergillus* in abundance. The concentration of this genus was 360/m^3 on the day before, 595/m^3 during and 525/m^3 on the day after mela. This fungus made up 52.63% of the total fungal isolates on the day before, 53.76% during and 74.58% after mela. Among the species identified, *A. luchuensis* emerged as the dominant one in the three samples. *A. flavus* was also present in the three samples. But, *A. niger* appeared during and after mela, *A. flavipes, A. violaceo-fuscus* and *A. fumigatus* before and during the mela. *Penicillium* was trapped before and during mela and the respective concentrations were 120/m^3 and 238/m^3. within this genus *P. fellutanum, P. frequentans, P. lansosum* and *P. oxalicum* were present before and during mela, while *P. citrinum* appeared only on the day before mela. *P. italicum, P. purpurogenum* and *P. granulaum* were isolated only on the day of the mela. Yeasts appeared in the three samples and the concentrations before, during and after the mela were 60/m^3, 140/m^3 and 10/m^3 respectively. *Curvularia indica* and *Mucos hiemalis* were obtained in the three days of sampling, *Drechslera australiensis*, *Rhizopus nigricans* on the day of and the day after mela, *Fusarium moniliforme* on the day before and after mela, *Chaetomium globosum* and the non-sporulating fungi on the day before mela, *Veronaea apiculata*, *Stachybotrys atra* and *Paecilomyces varioti* on the day of mela, while *Absidia heterospora* was seen on the day after. Two petri dishes were exposed at the nose level of the people standing. Several of the isolates that appeared on these plates were common to those obtained on plates exposed in the Andersen sampler. In addition, these open Plates's revealed *Absidia* sp. *Alternaria dianthicola*, *Curvularia maculans*, *Bidenticula cannae*, *Stachybotrys atra* and *Trichothecium roseum*. A sample of soil collected from the site of the mela was plated. A number of fungi also seen in air samples were observed. These fungi were *Aspergillus luchuensis, A. flavipes, A. violaceo-fuscus, A. flavus, A. fumigatus, Penicillium oxallium, P. frequentans, Mucor* sp., *Rhizopus* sp., *Absidia* sp., and *Stachybotrys atra.*

DISCUSSION

The genus *Cladosporium*, an ubiquitous saprobe that grows on a variety of dead organic matter, is the dominant fungus reported in most aerobiological surveys, whether gravimetric or volumetric, conducted in different geographic regions of the world- Australia (Frey and Durie 1962; Derrick and McLennan 1963), New Zealand (Dye and Vernon 1952; Di Menna 1955), Israel (Barkai-Golan et al. 1977), south Africa (Ordman 1963), Nigeria (Dransfield 1966; Ugulana 1975), Jamaica (Meredith 1962), Canada (Pady and Kapica 1956), U.K. (Hyde 1969; and the references therein), Germany (Stix 1977), U.S. A. (Morrow et al. 1964 and the references therein), India (Sreeramulu 1967 and the references therein; Rati and Ramalingam 1976; Bhat and Rajasab 1988). Its proportion (ca 25%) relative to total catch presently reported is near the lower end of the range of 25-80% reported in the above studies.

Earlier studies on the fungal air spora of Visakhaptanam also reported the ubiquity of Cladosporium (Subba Reddi 1970; Atluri and Subba Reddi 1981). By contrast the present Andersen sampler study showed *Aspergillus* which contributed ca. 43 % to the total aeromycoflora, as the predominating fungus, while *Cladosporium* with ca. 25% contribution was relegated to the second place. The dominance of *Aspergillus* was also reported in open plate surveys from five other places in India - Kanpur (Ranjan et al. 1952), Calcutta (Chakraverty 1974), Madras (Vittal and Ponnusamy 1979), Bombay (Mandakini 1985) and Lucknow (Singh et al. 1980) and also from Philippines (Reyes et al. 1970). While Reyes et al. (1970), Vittal and Ponnusamy (1979) and Singh et al. (1980) noted *Cladosporium* as the second dominant fungus, Ranjan et al. (1952) found only a few colonies. The five places cited above and Visakhapatnam is industrial centers. It appears that in big cities and industrial areas in India, *Aspergillus* may prove to be the dominant component of the fungal air spora.

The present study helped to identify 145 fungal types against 41-56 recognized earlier by Subba Reddi (1970) and, Atluri and Subba Reddi (1981). However the obligate parasitic fungi represented by uredospores, oidium, etc., and several saprophytic types with larges spores including *Phaeotirchoconis, Corynespora, Tetraploa, Trichoconis, Cercospora*, etc., are not reflected in the spore spectrum now revealed. The discrepancies evident in the relative frequencies of small spored and large spored fungi between the present and previous studies could be explained taking into consideration the relative efficiencies of the two sampling devices. The Andersen sampler collects 6 graded fractions of Petri dishes by a system of multi jet cascade impaction (Andersen 1958). May (1964) reported that stages I and II of the sampler show large differences in impaction performance and the intake efficiency for larges particles (> 12 μm in diameter) is poor. The cone over the top stage I is like an inverted funnel and directs a broad jet of air at the central zone of this stage causing impaction and subsequent large percentage loss of any large particles on its upper surface. With an aerosol of 20 μm diameter, the loss was estimated as 87%. As the particle diameter decreases, impaction becomes more efficient. May (1964) and Lidwell and Noble (1965) proved that stage III and below in the standard model of the sampler function in accordance with the data of Andersen (1958). Gregory (1973) also suggests that for particles larger than 8-10 μm, the addition of a pre-stage is desirable to avoid losses on the first stage.

The conidia of most species of *Aspergillus, Cladosporium* and *Penicillium* are less than 8μm in diameter and are equally impacted on stages III and IV. The diameter range of particles for impaction on these 2 stages is 2-6 μm (Andersen 1958). Thus, the present study the small spored fungi such as *Aspergillus, Cladosporium* and *Penicillium* are revealed well by the Andersen sampler. The wind impaction sampler employed in the previous studies is sensitive to wind speeds and particles size, with larger particles being impacted more efficiently than the smaller ones (Gregory 1973; Ogden et al. 1974). Therefore, the large spored fungi such as *Alternaria, Curvularia, Nigrospora, Periconia* etc., were revealed well by the previous studies. Based on the comparison of results obtained by the previous studies and those of the present, it is suggested that the 2 methods of study supplement each other data, and a combination of both methods is desirable in an aerobiological study to reveal as complete a picture of airspora as possible. The ascospores and basidiospores could not be revealed by Andersen sampler but it is likely that some of the mycelia sterilia might belong to these fungal groups.

A useful comparison can be made of composition and relative frequency of the species of *Aspergillus, Penicillium* and *Cladosporium* reported from India and elsewhere. Altogether there are 70 different species of *Aspergillus* as reported from 15 sampling stations, 10 located within and 5 outside India. The number of species isolated at different locations were: Mysore 37, Visakhapatnam 21, Gorakhpur 27, Delhi 17, Madras and Kanpur 9 each, Calcutta and Guntur 7 each, Modinagar and Egypt 5 each, Kuwait 19, Israel 22, Kansas (USA) 27, Cambridge 14. As was claimed by the investigators (Rati and Ramalingam 1976), the large number of species recorded at Mysore was because of the large amount of air sampled.

The study shows that *A. niger, A. nidulans, A. terries, A. fumigatus* and *A. flavus* have wider distribution. However, these are not the quantitatively more abundant at each of the locations studies. Thus at Visakhapatnam, *A. luchuensis* dominated; this together with *A. violaceo-fuscus, A. niger* comprised a major percentage of the genus (75.98). At Mysore *A. nidulans, A. ruber, A. awamori, A. flavus, A. amstelodami, A. versicolor* collectively made up the bulk of *Aspergillus* concentrations (Rati and Ramalingam 1976). At Cambridge *A. fumigatus, A. repens, A. versicolor* and *A. amstelodami* formed a major percentage (Hudson 1969) and at Kansas *A. niger* and *A. amstelodami* were the commonest species (Kramer et al. 1960a). Thus it is clear that the widely distributed species need not be abundant throughout their distribution range.

With *Penicillium*, there are altogether 62 species isolated from air samples collected in India and abroad. None of these species occurred at all the 8 stations. Two species viz., *P. frequentans* and *P. notatum* were present at 5 stations another 3 species that included *P. chrysogenum, P. oxalicum* and *P. rubrum* at 4 of the 8 stations. Each station appears to have its own dominant species. Thus, *P. fellutanum* and *P. rubrum* dominated the air of Visakhapatnam; *P. miczynskii* of Calcutta; *P. nigricans* and *P. citrinum* of Madras; *P. notatum* of Modinagar; *P. islandicum. P. thomii, P. variabile* and *P. varians* of Kuwait; and *P. oxalicum* of Kansas.

Within the genus *Cladosporium, C. herbarum* and *C. cladosporioides* appear to have a relatively wider distribution than any of the 11 species reported from 10 stations with in India and outside. These are also the dominant species at several stations; other dominant species included *C. sphaerospermum* at Madras and Cardiff, and *C. tenuissium* at Shillong.

In spite of restricted sampling the study showed the existence of a phenological periodicity in the incidence of total viable fungi. The period from October to March registered a higher frequency, with peaks occurring at any time between mid November and mid February. This observed periodicity agrees well with the pattern reported earlier at Visakhapatnam through day to day analysis by visual method (Atluri and Subba Reddi 1981).

The year round presence of fungal spores in the atmosphere has been reported the world over, however the phenological periodicity appears to vary between temperate and tropical regions of the world. In England and Kansas (USA) experiencing a temperate climate with clear cut seasons and with the winter being severe, the number of spores is the highest in the summer (growing season) and the lowest during the winter (Adams 1964; Hyde 1969; Kramer 1974). On the other hand, in the tropical regions the climatic seasons are less clearly defined and in some areas there is much overlapping and the winters are relatively mild. Under such tropical conditions, some workers found elevated concentrations of spores during the winter (Sreeramulu and Ramalingam 1966; Mishra and Kamal 1971; Ramalingam 1971; Bajaj 1978; Chen et al. 1978; Mallaiah and Rao 1980; Atluri and Subba Reddi 1981; Bhat and Rajasab 1988); some observed peaks of almost equal magnitude during winter as well as summer (Rajan et al. 1952; Barat and Das 1963; Agarwal et al. 1969; Wadhwani et al. 1979; Singh et al. 1980; Gaur and Kasana 1981) and some reported peaks during the transition between winter and summer (Moustafa and Kamel 1976; Tuner 1966). On the other hand, winter recorded the least count and the rest of the year showed high incidence (Barkai-Golan 1958; Barkai-Golan and Glazer 1962; Barkai-Golan et al. 1977) and in the northern Nigeria, peak was obtained at the beginning of the dry season (October –November) and again during the first half of the rainy season between April and mid July (Dransfield 1966). Still others reported a large summer maximum and a small winter maximum over agro-environments and the peaks were related to the harvesting of the crops (Subba Reddi 1970; Varma 1986).

As to the key factors controlling the fungal growth, reproduction and conidial dissemination, different authors report different results. Thus, Turner (1966) from Hong Kong observed an overall inverse relationship between total colonies and temperature and also with the rainfall. Derrick and McLennan (1963) from Melbourne (Australia) stated that high temperature favoured total fungal spore production and dissemination but production was less under dry conditions. Amin and Bokhari (1979) reported a large peak in July, the hottest and driest month in Shiraz; evidently there is a correlation between dryness of the air and the incidence of the spores. Wadhwani (1979) from Lucknow related the peaks during October and again during early summer (February to April) to the harvesting of crops and to the relatively higher temperature which she states as more suitable to the growth of fungi. Mishra and Kamal (1971 from Gorakhpur stated that during the winter when the peaks were recorded, heavy dew accumulation during the nights provided enough moisture for the germination and multiplication of the fungal spores. Kramer (1974) explained how wet periods provided excellent growing conditions for the fungi in the surrounding areas of Kansas (USA), resulting in extremely high number of spores being released into the atmosphere throughout the growing season. Cammack (1955) and Dransfield (1966) from Africa argued that the availability of suitable substrate and moisture are the cardinal factors limiting the fungal growth. Mallaiah and Rao (1980) attempted to explain the difference in the seasonality of conidial fungi between the temperate and tropical regions by referring to Cochrane (1958) who stated that 20-30ºC and 25ºC are the optimum temperatures for fungal growth and plant parasitizing fungal respectively. These authors maintain that such temperatures are realized during the winter in the tropical regions and during summer in the temperate regions. The realization of summer maxima in some of the localities in the tropical regions (Subba Reddi 1970; Varma 1986) does not agree with the conclusion of Mallaiah and Rao (1980).

The genus *Aspergilllus* is reported in some localities to have no definite seasonal periodicity, but in some localities it does exhibit a seasonal trend in its occurrence. Thus, at Delhi (India) (Agarwal et al. 1969) and Israel (Barkai-Golan et al. 1977), its periodicity was less obvious. Also at Cambridge and Nottingham (UK) the periodicity was less obvious (Pawsey and Heath 1964; Hudson 1969). At Kansas (USA), there was a seasonal trend with high numbers in spring and fall and low numbers during summer (Kramer et al. 1960a). There *A. amstelodami*, the dominant species, was somewhat more abundant during the growing season, and *A. niger* the next dominant occurred more or less at the same frequency throughout the year. In Kuwait Moustafa and Kamal (1976) observed a periodicity with higher colony counts during November-December and during February-March. In Hong Kong Turner (1966) found higher numbers during September-December and again in June. At Gorakhpur, *A. syndowi, A. terreus*, *A. flavus, A. niger* were widely distributed throughout the year, while *A. fischeri, A. fumigatus, A. nidulans* and *A. variecolor* were dominant during summer (Mishra and Kamal 1971; Mishra 1972). Rati and Ramalingam (1976) observed at Mysore winter maxima for the commonest species like *A. nidulans*, *A. niger, A. awamori, A. flavus, A. amstelodami* and peaks both in winter and summer seasons for *A. candidus, A. bruneo-uniseriatus* and *A. ochraceous*. Wadhwani (1979) at Lucknow observed higher incidence during February-April. Bajaj (1978) found at an agricultural site in Nagpur higher counts in May, August and December during 1971 and during the summer of 1972, whereas at the residential site he recorded higher counts during winter, with maximum during February–March. Chakraverty (1974) from Calcutta stated that the colonies were comparatively more during summer and rainy seasons, and less during winter. Barat and Das (1963) reported high counts of *A. niger* during summer. The present results agree with those of Rati and Ramalingam (1976) in that at both localities a winter maxima was obtained for the genus in general and, different species exhibited different peaks. It is to be noted that some authors reported increased incidence of *Aspergillus* following occasional rains during summer and the peak reported in the present study during May agrees well with Bhat and Rajasab (1988), and Bajaj (1978).

Compared to *Aspergillus,* there have been limited studies on *Penicillium* in the air. Pawsey and Heath (1964) from Nottingham (UK) reported a less clearly defined periodicity, with higher numbers generally occurring during winter and spring. At Kansas (USA) Kramer et al. (1960a) observed this fungus showing a definite seasonal trend despite a great day to day variation caused by rainfall. The highest numbers of colonies have occurred from July to November or December and lowest in late winter and early spring. In India, Wadhwani (1979) from Lucknow, reported a winter peak and Chakraverty (1974) observed at Calcutta a higher frequency during summer as well as rainy seasons. The pattern observed in the present study agrees with that reported at Calcutta.

With regard to the environmental factors influencing the incidence of *Aspergillus* and *Penicillium*, different authors have given different explanations and these are mentioned in the pages that follow. Obviously there is much need for work in this direction which will ultimately help in formulating the models to predict the concentrations of the spores of these fungi in the atmosphere. *Cladosporium* is known in various places in the world as a distinct seasonal fungus and its periodicity, in general is reflected in the periodicity of total spores in the air. Concerning the factors that determine its incidence in the air, Hirst (1953) and, Gregory and Hirst (1957) recorded abundant spore incidence during high temperatures and high humidity. Summarizing results of 8 years, Adams (1964) concluded that a mean temperature of over 60°F and little or no rain seems to be essential for mass production and liberation of spores. Ainsworth (1952) recorded that the number of colonies on plates exposed during rain greatly exceeded that occurring during dry periods. Davies (1969) demonstrated experimentally the effect of rain in dislodging spores from conidiophores. Wind tunnel and slit sampler experiments of Harvey (1967) showed that the air chares with water droplets removes many more spores from cultures than does the dry air. Davies (1959) mentioned that an accumulated temperature below 20°C limits the growth of *Cladosporium.* Derrick and McLennan (1963) stated that the growth is limited by both dryness and low temperature, but responds rapidly to warmer temperatures in association with adequate moisture. Chakraverty and Nandi (1972) found that this fungus thrives at a temperature range of 22-30°C such correlation is also observed in the present study. However, it appears that this relationship is not much exacting since higher catches are sometimes maintained in the face of an increase in temperature, particularly following rains during the hottest months of May and June (Sreeramulu and Ramalingam 1964; Mallaiah and Rao

1980; Atluri and Subba Reddi 1981; Varma 1986). Probably, the substrate and moisture availability might play a greater role as emphasized by Dransfield (1966). During the period the fungus was abundant, there was a greater accumulation of vegetable matter (particularly of dicot leaves) and heavy dew which favoured abundant growth and mass production of spores. Microscopic examination of such substrata under heavy dew formation revealed *Cladosporium* as the most common type (Sreeramulu and Ramalingam 1964). The observation of Richards (1956), Hyde (1969) and Bajaj (1978) that catches at agricultural areas are higher than those at residential ones implies the significance of substrate. The association of abundance of this fungus and the accumulation of dead and decaying vegetable matter was also observed by Sreeramulu and Ramalingam (1966), Davies (1969) and Singh et al. (1981). The low temperatures in the winter of the temperate regions might limit fungal growth but the relatively higher temperatures of the summer of the tropical regions might not inhibit the growth of most of tropical fungi. But it is the scarcity of the availability of suitable substrate and moisture that hinders the fungal growth and developments in the tropics. The increase in the incidence of *Alternaria, Aspergillus, Cladosporium, Curvularia* and *Nigrospora* following the rains during may and June, the hottest months of the year, underlines the significance of moisture (Sreeramulu and Ramalingam 1964; Bajaj 1978; Mallaiah and Rao 1980; Atluri and Subba Reddi 1981; Bhat and Rajasahab 1988). Even in the temperate regions the promotive effect of moisture on spore incidence during the growing season was recognized (Kramer 1974).

Unlike other environments, the atmosphere contains only the allochthonous microorganisms. As to the origin of fungal spores in the air, there are two conflicting views. Pady (1957) considered that most spores come from the soil fungal flora, whereas Gregory (1961) argued that they are mainly derived from the fungi growing on vegetation. Comparing aero and soil mycoflora studied simultaneously, Dransfield (1966) concluded that it is improbable that the soil makes more than a small contribution to the fungal population of the atmosphere. Mishra and Kamal (1971) expressed that both soil and vegetation are equally important as parent stock for the air spores. I believe in giving equal importance to both soil and vegetation as sources for the fungal air spora. Considering the prevalence of Aspergilli and Penicillia over other fungi in the soil (Behera and Mukerji 1985 and the references therein), it may be assumed that the soil may well be a major source of their incidence in outdoor air, then their pattern of relative dominance in the air should reflect that in the soil. While reviewing the distribution and detection of root disease fungi, Warcup (1959) stated that species of *Penicillium* appear common in temperate soils while those of *Aspergillus* in tropical soils (see also Behera and Mukerji 1985, and the references therein). Rati and Ramalingam (1976) compared their own results from Mysore (India) with those of Hudson (1969) from Cambridge (England), both used Andersen sampler. It was shown that Aspergilli predominated over Penicillia in the tropical air while the reverse is true in the temperate regions. To substantiate their conclusion they also cited the results of open plate method from India (Sandhu et al. 1964; Agarwal et al. 1969). Recent cultural studies (open plate) from India also proved the predominance of Aspergilli over Pencillia (Viattla and Ponnusamy 1979; Mandakini 1985; Bhat and Rajasab 1988). In the present volumetric study also the frequency of *Aspergillus* exceeded that of *Penicillium* by more than four times. Studies from other tropical regions of the world are not entirely in conformity with the predicted abundance of *Aspergillus* over *Penicillium*. While *Aspergillus* with 17.1% contribution to the total colonies was dominating over *Penicillium* (14.0%) at Kuwait (Moustafa and Kamal 1976), it was inferior with 10.5% to *Penicillium* with 17.1% contribution at Hong Kong (Turner 1966). In Israel both open plate (Barkai-Golan 1958; Barkai Golan and Glazer 1962) and volumetric methods (Barkai-Golan et al. 1977) established the predominance of *Penicillium* over *Aspergillus*. The same was the situation at Samaru, Nigeria (Dransfield 1966). Obviously more studies preferably volumetric spread over different regions of the tropics are required to test the validity of the above prediction. Not only the volumetric study of Hudson (1969), other open plate studies from England (Richards 1954; Pawsey and Heath 1964) also witnessed the superiority of Penicillia over Aspergilli. The volumetric study performed with Pady Rittis and General electric samplers at Kansas (USA) by Kramer et al. (1960a) and open plate surveys from New Zealand (Di Menna 1955) and Australia (Derrick and McLennan 1963) showed a slight superiority of *Penicillium* over *Aspergillus*. Then, the question arises whether the temperature controls the relative abundance of *Aspergillus* and *Penicillium*. Available literature in this respect is somewhat inconsistent.

The term allergy was coined by Von Pirquet in 1906 and has been defined as the acquired, specific altered capacity to reach (see Agarwal and Shivpuri 1974). It is acquired only by exposure to an allergen. It is specific to that allergen and it is characterized by a reaction not present before sensitization (Lacey et al. 1972). According to Pepys (1969), the allergic reaction may be immediate conforming to Type I reaction to be delayed several hours after exposure to allergen taking the form of Type III reaction. People may be grouped into 2 broad categories according to their reaction of allergens. One group i.e., atopic with 10% of the population in which Type I reactions occurs. These people are easily sensitized and are therefore considered to be predisposed to allergy; this group develops non precipitating antibodies (regains, immunoglobin IgE). The other 90% of the population are non-atopic in whom Type III reaction may be produced. This reaction by contrast requires much longer or greater exposure for sensitization. This group produces precipitating antibodies IgG or IgM (see Gregory 1973; Lacey 1975).

About the allergenic nature of fungal spores, it was Blackey (see Chanda 1970) who provided a clue when he showed that bronchial catarrh could be induced by the inhalation of *Penicillium* and *Chaetomium* spores. However, it was not until after Feinberg's report in 1935 that fungi were considered to be an important cause of allergy. Since then several investigators provided ample evidence as to the allergic significance of fungal spores. The achievements of the extensive studies made in this field have vividly reviewed from time to time (Maunsell 1954; Werff 1958; Blatt 1962; Agarwal and Shivpuri 1974; Shivpuri 1982). The present notion is that all fungal spores must be considered potentially allergic (Gregory 1973) and among all major groups of fungi are genera which produce spores causing inhalant allergy (Southworth 1974). However, some genera are more important (allergenically) than others and a rather limited number are involved in skin testing and therapy of allergic individuals (Lewis et al. 1975; Singh et al. 1979; Gravesen 1979; Tilak et al. 1980; Shivpuri 1982; Lopez and Salvaggio 1985). Although Targow (1966a,b) maintains that there is no species specific antigenicity among fungi, there is confirmatory evidence to show that species differ in their allergenicity (Agarwal et al. 1968 and the references therein) and this warrants species identification of fungi present in the atmosphere (Austwick 1966). The present study fulfilled its objective by segregating the 'aspergilli' component of the air spora reported by Subba Reddi (1970) and Atluri and Subba Reddi (1981) into *Aspergillus* and *Penicillium* and by providing species differentiation of each genus and also of *Cladosporum* which are established to be the universal dominants of the air spora (Prince and Meyer 1976).

In figure 5 of Andersen (1958) particles larger than 5 μm are shown to get deposited on stages I & II of the sampler which correspond to nose (also pharynx). Such particles in the study include *Curvularia, Drechslera* and *Alternaria*. Their deposition in the upper respiratory tract produce reactions resulting in rhinitis and bronchial asthma. The spores of *Cladosporium, Penicillium* and *Aspergillus* were also evident on stages I and II which could be possible because their spores occur in clumps and singly and hence could cause both kinds of reactions and allergic manifestations.

Spores associated with Type I allergy are usually larges than 5 μm while those associated with Type II are usually smaller (see Lacey 1975). Natural spore concentrations may cause Type I allergies. During 2 year study period, the spore concentrations obtained per sample varied from a low of $312/m^3$ to a high of $4855/m^3$. On the other hand, type III allergy results from concentration over $10/m^3$. As many as $3.3 \times 10/m^3$ have been recorded during the harvesting of bajra

(*Pennisetum americana*) in the suburbs of Visakhapatnam (Sankara Rao 1988). Prolonged exposure to smaller concentration may cause insidious development of diseases (Pepys 1969). A consideration of the distribution of spores of different fungi on different stages of Andersen Sampler suggested that *Alternaria, Curvularia* and *Drechslera* and other spore types with large spores may cause Type I allergy while *Aspergillus, Penicillium* and *Cladosporium* and others with small spores, dispersed in clumps or singly, may produce both forms of allergy.

The study suggested that the fungi are to be treated as important and pernicious bio-pollutants of aerial environment of Visakhapatnam. In addition the city receives in large quantity of chemical pollutants from industries such as Bharat Heavy Plate and Vessels, Hindustan Zinc Smelter, Coromandel Fertilizers, Hindustan Petrochemicals and Steel Industry (Vittal Murty 1981-84). It is known that chemical and biogenic air pollutants interact with each other (Newark 1970; Nilsson and Nybom 1978). It is therefore likely that such a situation might be happening at Visakhapatnam resulting in synergistic or reinforcing effects, which in turn exacerbate the discomfort of allergy victims. Therefore the allergy clinicians and environmentalists should take into consideration the possibility of synergism between chemical and bio-pollutants in their plans to manage allergies.

REFERENCES

Abdel-Hafez, S.I.I., Shoriet, A.A.M., Abdel-Hafez, A.I.I. and El-Maghraby, O.M.O. 1986. Mycoflora and mycotoxin producing fungi of air dust particles from Egypt. Mycopathogia 93: 25-32.

Adams, K. F.1964. Year to year variations in fungus spore content of the atmosphere. Acta Allergol. 19: 11-50.

Agarwal, M.K. and Shivpuri, D.N. 1974. Fungal spores- their role in respiratory allergy. Adv. Pollen spore Res. 1: 78-128.

Agarwal, M.K., Shivpuri, D.N. and Mukerji, K.G. 1969. Studies on the allergic fungal spores of the Delhi, India. Metropolitan area. J. Allergy 44: 193-203.

Agarwal, M.K., Mukerji, K.G., Shivpuri, D.N. and Shali, P.L. 1968. Species specific antigens in fungi. Asp. Allergy Appl. Immunol. 2: 69 (cited from Agarwal and Shivpuri 1974).

Ainsworth, G.C. 1952. Incidence of air borne *Cladosporium* spores in the London region. J. Gen. Microbiol. 7: 358-361.

Airy, H. 1874. Microscopic examination of air. Nature (London) 9: 439-440.

Al-doory, Y. 1967. Further studies of the fungal flora of the air in San Antonio, Texas. J. Allergy 40: 145-150.

Al-doory, Y. 1970. Application of Andersen sampler in studying air borne fungi in San Antonio,

Texas. Mycopathol. Mycol. Appl. 42:293-298.

Amin, R. and Bokhari, M.H. 1979. Survey of atmospheric fungus spores in Shiraz, Iran. Ann. Allergy 42: 246-247.

Andersen, A.A. 1958. New sampler for the collection, sizing and enumeration of viable air borne particles. J. Bacteriol. 76: 471-484.

Atluri, J.B. and Subba Reddi, C. 1981. The fungus airspora of Visakhapatnam. Proc. Indian Nat. Acad. 47:731-747.

Atluri, J. B. and Subba Reddi, C. 1982. Periodicity in the air borne ascospores of Pleosporea. Indian Phytopahtol. 35: 504-506.

Austwick, P.K.C.1966. The role of spores in allergies and mycoses of amn and animals. pp. 321-337. In: M. F. Madelin (Ed.). The fungus spore, Butterworths, London.

Bajaj, A. 1978. Study of viable spores in air at two different sites of Nagpur. J. Palynol. 14:136-149.

Barat, R. and Das, A.C. 1963. Fungi of allergic significance in air over Calcutta. Ann. Biochem. 23:461-466.

Barkai-Golan, R. 1958. A stud y of airborne fungi in Israel. Bull. Res. Coun. Israel 60: 247-258.

Barkai-Golan, R. and Glaze, I. 1962. Airborne fungi in Eilat and Tel- Hashomer, Israel. J. Allergy 33: 342-348.

Barkai-Golan, R., Frank, M., Kantor, D., Karadavid, R. and Toshner, D. 1977.. Atmospheric fungi in the desert town of Arad and in the coastal plain of Israel. Ann. Allergy 38: 270-274.

Barlow, H.S. 1963. A collaborative study of airborne molds in southern and central Ontario. Ann. Allergy 21: 569-576.

Bartzakas, C.A. 1975. Relationship between the meterological conditions and the air borne fungal flora of the Athens metropolitan area. Mycopathologia 57: 35-38.

Baruah, H.K. and Chetia, M. 1966. Aerospora and allergenic human diseases: a study of certain fungal spores and pollen grains of Gauhati. Indian J. Exp. Biol. 4: 236-238.

Baxter, L.J. and Cookson, J.T. 1983. Natural atmospheric microbial conditions in a typical suburban area. Appl. Environ Microbiol. 45:919-934.

Behera, N. and Mukerji, K.G. 1985. Seasonal variation and distribution of microfungi in forest soils in Delhi. Folia Geobotanica et Phytotaxonomica 20: 291-312.

Bernstein, T.B. and Feinberg, S.M. 1942. Airborne fungus spores. A five year survey of daily mold spore content of Chicago air. J. Allergy 13: 231-241.

Bhaskara Rao, P. and Mallaiah, K.V. 1985. Viable and culturable fungi in the air over Nagarjunasagar. Indian Bot. Reptr. 4: 55-57.

Bhat, M. and Rajasab, A. H. 1988. Incidence of viable fungal airspora in Gulbarga during 1984-86: A two years survey. Indian J. Aerobiol. 1: 10-16.

Bhati, H. S. and Gaur, R.D. 1979. Studies on aerobiology-atmospheric fungal spores. New Phytologist 82: 519-527.

Bhatt, H. 1962. Microbial allergy- a critical review. Ann. Allergy. 20: 263-281.

Cadham, F.T. 1924. Asthma due to grain rusts. J. Am. Med. Ass. 83: 27.

Calvo, M.A., Guarro, J., Suarez, G. and Ramirez, C. 1980. Airborne fungi in the air of Barcelona, Spain. Ann. Allergy 45: 115-116.

Cammack, R.H. 1955. Seasonal changes in three common constituents of the airspora of southern Nigeria. Nature (London) 179: 1270-1272.

Chakraverty, R. 1974. Fungi in air and human allergy. J. Sci. Club. 28: 83-106.

Chakraverty, R. 1976. Aeromycology of the Indian botanic garden (Calcutta). Asp. Allergy Appl. Immunol. 9: 151-157.

Chakraverty, R. and Nandi, P. 1972. The seasonal periodicity of *Cladosporium,* a common allergen in air over Calcutta and suburbs. Tans. Bose Res. Inst. 35: 45-50.

Chanda, S. 1970. An old and rare Indian publication. Sci. Cult. 36: 391-392.

Chandwani, G.H., Balakrishnan, M.S. and Padmanabhan, S.Y. 1963. *Helminthosporium* disease of rice. A study of the spore population of *Helminthosporium oryzae* over rice fields. J. Indian Bot. Soc. 42: 1-14.

Chaubal, P.D. and Kotmire, S.Y. 1982. Aerobiological studies at Kolhapur, India Acta Bot. Indica 10: 100-102.

Chen, Z.C., Hsiung, Y.M and Tseng, H.Y. 1978. Seasonal variation of fungal air spora Taipei. Pp. 148-155. In: Proc. 1st international conference of aerobiology, Munich.

Cochrane, V.W. 1958. Physiology of fungi. John Wiley, New York, London.

Collins-Williams, S. C. and Best, C.A. 1955. Atmospheric mold counts in Toronto, Canada. J. Allergy 26: 468-470.

Collins-Williams, C., Kuo, H.K., Garey, D.N., Davidson, S., Collins-Williams, D., Fitch, M. and Fisher, J.B. 1973. Atmospheric mold counts in Toronto, Canada. 1971. Ann. Allergy 31: 69-71.

Cunningham, D.D. 1873. Microscopic examinations of air. Government Printers, Calcutta.

Davies, R.R. 1959. Detachment of conidia by cloud droplets. Nature (London) 183: 1695.

Davies, R.R. 1969. Spore concentrations in the atmosphere at Ahmadi, a new town in Kuwait. J. Gen. Microbiol. 55: 425-432.

De Bary, Y. A. 1853. Lintersuchugne uber die Brandpilze und die durch sie verusachten krankheiten der pflanzen mit Rucksicht auf das Getreide und andere Nutpflanzen.

Derrick, E. and McLennan, E.I. 1963. Fungus spores found in the air in Melbourne (Victoria), Australia. Acta Allergol. 18: 26-43.

Dimmick, R. L. and Akers, A.B. (Eds.) 1969. An introduction to experimental aerobiology. Wiley-Interscience, New York, London, Surrey and Toronto.

Di Menna, M. E. 1955. A quantitative study of airborne fungus spores in Dunedin, New Zealand, Trans. BR. Mycol. Soc. 38: 119-129.

Dixit, R.B and Gupta, J. S. 1981. Seasonal and diurnal census of air spora over a barley field. Proc. Nat. Acad. Sci. B51: 44-46.

Dransfield, M. 1966. The fungal air spora at Samaru, northern Nigeria. Trans. Br. Mycol. Soc.49: 121-132.

Durham, O. C. 1942. Airborne fungus spores as allergies. In: Moultan, F.R. (Ed.) Aerobiology. Am. Ass. Adv. Sci. Symposium 17.

Dye, M. H. and Vernon, T. R. 1952. Airborne mould spores, N. Z. J. Sci. Technol. 34: 118-127.

Edmonds, R.L. (Ed.) aerobiology: the ecological systems approach. Dowden, Hutchinson and Ross, Inc. Strousburg/Pennsylvania.

Edmonds, R.L. and Benninghoff, W.S. 1979. The challenges of aerobiology: sampling instruments. In: Ed. R.L. Edmonds, Aerobiology: The ecological systems approach. pp. 377-379, Dowden Hutchinson & Ross, Inc., Strousburg/Pennsylvania.

Ehrenberg, C.G. 1849. Passat-staub und blut-regen, Berlin.

Ellis, M.B 1971. Dematiceous Hyphomycetes. Commonwealth Mycological Institute, Kew, England.

Ellis, M.B. 1976. More Dematiaceous Hyphomycetes. Commonwealth Mycological Institute, Kew, England.

Feinberg, S.N., Durham, O.C. and Dragstedt, C.A. (Ed.) 1946. Allergy in practice. Year Book Publishers, Chicago.

Frey, D. and Durie, E. B. 1960. The incidence of airborne fungi in Sydney. Mycopath. Den. Hagg. 13: 93-99.

Frey, D. and Durie, E. B. 1962. Estimation of airborne fungus spores: a comparison of slide and culture methods. Mycopath. Mycol. Appl. 16: 295-303.

Gaur, R.D. and Kasana, M.S. 1981. Studies on aerobiology of Modinagar. J. Indian Bot. Soc. 60: 266-277.

Gleditsch, I.G. 1766. Uber den luftstaub, in absccht auf die darunter befindlerhin scamen von eningen arten der schwamme und Erdflexhten. Gleditsch Physie, Bot. Deconum Abh. Hulle II: 323-344.

Gopi. T.V. 1984. Incidence of fungal spores in the air over Calicut university area in the year 1981. Abst. SV-20, Fourth Indian Palynological Conference, March 19-21, 1984, Andhra University, Waltair.

Gravesen, S. 1979. Fungi as a cause of allergic disease. Allergy 34: 135.

Gregory, P.H. 1952a. Fungus spores. Trans. Br. Mycol. Soc. 35: 1-18.

Gregory, P.H. 1952b. Spore content of the atmosphere near the ground. Nature (London) 170: 475.

Gregory, P.H. 1961. The microbiology of the atmosphere. Leonard Hill. London.

Gregory, P.H. 1973. The microbiology of the atmosphere. 2nd Ed. Leonard Hill, London.

Gregory, P. H. and Hirst, J. M. 1957. The summer air spora at Rothamsted in 1952. J. Gen. Microbiol. 17: 135-152.

Gregory, P.H. and Monteith, J.L. (Eds.) 1967. Airborne microbes, Sym. Soc. Gen. Microbiol. 17 Cambridge University Press.

Grose, E.S., Maria, S. and Neily, M. 1967. Air borne fungus spores in Bogota, Colombia: A Five year study. J. Int. Sco. Hum. Anim. Mycol. 6: 42-50.

Harvey, R. 1967. Airspora studies at Cardiff. I. *Cladosporium*. Trans. Br. Mycol. Soc. 50: 479-495.

Hirst, J. M. 1953. Changes in atmospheric spore content: diurnal periodicity and the effects of weather. Trans. Br. Mycol. Soc. 36: 375-93.

Hudson, H. J. 1969. Aspergilli in the air spora at Cambridge. Trans. Br. Mycol. Soc. 52: 153-159.

Hyde, H.A. 1969. Aeropalynology in Britain: an outline. New Phytologist 68: 579-590.

Ingold, C. T. 1971. Fungal spores: their liberation and dispersal. Clarendon Press, Oxford.

Jayaprakash, K.B. and Ramalingam, A. 1981. *Aspergillus fumigatus* in the air of working environments at Mysore. Indian J. Bot. 4: 17-23.

Jayaprakash, K.B., Rati, E. and Ramalingam, A. 1978. *Aspergillus flavus* in the air of working environment. Curr. Sci. 47: 920-921.

Kalra, S. I. and Dumbry, D.G. 1957. Aerobiology of army medical campus, Poona. Armed Forces Med. J. (India) 13: 3-16.

KONGER, G. and Baruah, H.K. 1958. Incidence of air borne spores in the potato plantations of upper shillong. J. Univ. Gauhati. 9: 81-89.

Kotmire, S.Y. and Chaubal, P.D. 1984. Aeromycoflora at Gadhinglaj Abst. SV. 17. Fourth Indian Palynological Conference, March 1-21, 1984, Andhra Unversity, Waltair.

Kramer, C.L. 1974. Seasonality of air borne fungi. Pp. 415-424. In: H. Leith (ed.) phenology and seasonality modeling. Springer-Verlag, New York.

Kramer, C.L. and Pady, S. M. 1960. Kansas aeromycology. XI: Fungi Imperfecti, Trans. Kans. Acad. Sci. 63: 228-238.

Kramer, C. L, Pady, S. M. and Rogerson, C.T. 1960a. Kansas aeromycology V: *Penicillium* and *Aspergillus*. Mycologia LII (4): 545-551.

Kramer, C.L, Pady, S.M. and Rogerson, C. T. 1960b. Kansas aeromycology VIII: Phycomycetes. Trans. Kans. Acad. Sci. 63: 19-23.

Kramer, C.L., Pady, S. M., Rogerson, C.T. and Ouye, L. 1959a. Kansas aeromycology II. Materials methods and general results. Trans. Kans. Acad. Sci. 62: 184-199.

Kramer, C.L., Pady, S.M., Rogerson, C.T. and Ouye, L. 1959b. Kansas aeromycology III: *Cladosporium*. Trans. Kans. Acad. Sci. 62: 200-207.

Kulkarni, R.L. and Tilak, S.T. 1983. Atmospheric concentration of ascospores. Indian J. Bot. 6: 141-145.

Lacey, J. 1975. Potential hazards to animals and man from microorganisms in fodders and grain. Trans. Br. Mycol. Soc. 65: 171-184.

Lacey, J., Pepys, J. and Cross, T. 1972. Actinomycetes and fungus spores in air as respiratory allergens. D. Shapton and R. Board (Ed.). In: safety in Microbiology, Society of Applied Bacteriology Technical Series No. 6: 151-184.

Lewis, W. H., Imber, W.E. and Maniotis, J. 1975. Allergy epidemiology in the St. Lious, Missouri Area. I. Fungi. Ann. Allergy 34: 374-384.

Lidwell, O.M. and Noble, W.C. 1965. A modification of the Andersen Sampler for use in occupational environments. J. Appl. Bacteriol. 28: 280-282.

Liebeskind, A. 1965. Mold allergy in Haifa. Ann. Allergy 23: 158-161.

Long, R., Yin, R., He, H., Li, Z. and Liu, D. 1987. Molds in air in southern part of Chengdu (China). J. West. China Univ. Med. Sci. 18: 60-63.

Lopez, M. and Salvaggio, J.E. 1985. Mold sensitive asthma. Clin. Rev. Allergy 3: 183-196.

Lu, Y.L.C., Tzeng, J. C. and Huang, S. G. 1969. Airborne fungi in Taipei city. Chi. J. Microbiol. 2: 102-111.

Mallaiah, K.V. and Rao, A.S. 1980. Airspora of groundnut fields. Proc. Indian Acad. Sci. (Plant Sci.) 89: 269-281.

Mallaiah, K.V. and Rao, A.S. 1982. Aerial dissemination of uredinospores of groundnut rust. Trans. Br. Mycol. Soc. 78: 21-28.

Mandakini, S. 1985. Seasonal variation and vertical stratification of the atmospheric fungal flora in Bombay. Abst. 3rd Natl. Conf. Aerobiology, Kalyan (Bombay), India.

Maunsell, K. 1954. Respiratory allergy to fungus spores. Progr. Allergy 4: 457-520.

May, K. R. 1964. Calibration of a modified Andersen bacterial aerosol sampler. Appl. Microbiol 12: 37-43.

Mehta, K. C. 1933. Rusts of wheat and barley in India. Indian J. Agric. Sci. 3: 939-962.

Meredith, D. S 1962. Some components of the air spora in Jamaican banana plantations. Ann. Appl. Biol. 50: 577-594.

Micheli, P.A. 1729. Nova plantarum genera.

Miquel, P. A. 1883. Les organisms vivants de 1 atmosphere. Paris.

Mishra, R.R. 1972. Aeromycology of Gorakhpur : IV. Periodical fluctuations of aerospora. Mycopath. Mycol. Appl. 48: 213-222.

Mishra, R.R. and Kamal. 1971. Aeromycology of Gorakhpur III. Seasonal variation in air fungal spora. Mycopath. Mycol. Appl. 45: 301-310.

Mishra, K.B. and Sahay, A. 1986. Suspended dust mycoflora of some public places during summer months at Gaya. Mendel 3: 160-163.

Mishra, K.B. and Sinha, A. 1985. Efficacy of the mixture of two nutrient media for isolating aerofungi, pp. 239-243.In: (ed.) Chandra, N. Atmospheric Biopollution. Environmental Publications, KARAD.

Mishra, K.B., Sinha, S.S. and Sahay, A. 1981. Aeromycological work at Gaya. I.A.S. News letter.

1: 3-4.

Morrow, M.B., Lowe, E.P. and Prince, H. E. 1942. Mold fungi in etiology of respiratory allergic diseases. 1. Survey of airborne spores. J. Allergy 13: 215.

Morrow, M. B., Meyer, G. H. and Prince, H.E. 1964. A summary of airborne mold surveys. Ann. Allergy 22: 575-587.

Moustafa, A.F. and Kamal, S. M. 1976. A study of fungal spore populations in the atmosphere of Kuwait. Mycopathologia 59: 29-35.

Muir, D.C.F. 1972. Deposition and clearance of inhaled particles. Pp. 1-20. In: Muir, D.C.F (ed.) Clinical aspects of inhaled particles. Heinemann. London.

Mukerji, K.G., Agarwal, M. K. and Saxena, A. S. 1968. Where from the allergic fungal spores are coming? pp. 181-189. In: Asp. Allergy Appl. Immunol. New Heights Publishers, Delhi.

Nemark, M. F. 1970. Recent developments in pollen aerobiology. Ann. Allergy 28: 149-152.

Nilssson, S. and Nybom, R. 1978. Particles on pollen. In: International Aerobiology News letter (Special issue) 8: 14-15.

Ogden, E.C., Raynor, G.S. Hayes, J.V., Lewis, M. D. and Haines, H. J. 1974. Manual for sampling air borne pollen. Hafner Press, New York.

Ogunlana, E. O. 1975. Fungal airspora at Ibadan, Nigeria. Appl. Microbiol. 29: 458-463.

Ordman, D. 1963. The air borne fungi in Johannesburg: a second five year survey 1955-1959. S. Afr. Med. J. 37: 325-328.

Padmanabham, S.Y., Ganguly, D. and Balakrishna, M.S. 1953. *Helminthosporium* disease of rice. II. Source and development of seedling infection. Indian Phytopathol. 6: 96-105.

Pady, S. M. 1954. An improved slit sampler for aerobiological investigations. Trans. Kans. Acad. Sci. 57: 157-163.

Pady, S. M. 1957. Quantitative studies of fungus spores in the air. Mycologia 69: 33-353.

Pady, S. M. and Kapica, L. 1956. Fungi in air masses over Montreal during 1950 and 1951. Canad. J. Bot. 34: 1-15.

Pady, S. M. and Kramer, C. L. 1960. Kansas aeromycology. X: Basidiomycetes. Trans. Kans. Acad. Sci. 63: 125-134.

Pandey, K.K. 1979. Variation in the air mycoflora over a fish market and a field. Acta Botanica Indica 7: 167-170.

Pasteur, L. 1861. Memoire sur les corpuscles organizes qui existent dans l atmosphere. Examen de la doctrine des generations spontanees. Ann. Sci. Na. (Zool.) 4 Ser.16: 5-98.

Pateria, A.K. and Sahu, J.R. 1982. An analysis of atmospheric fungal spores a sagar, India. Acta Bot. Indica. 10: 152-155.

Pawsey, R.G. and Heath, L.A. F. 1964. An investigation of the spore population of the air at Nottingham. Trans. Br. Mycol. Soc. 47: 351-355.

Pepys, J. 1969. Hypersensitivity diseases of the lungs due to fungi and organic dusts. Monographs in Allergy Vol. 4. S. Karger, Basle.

Prince, H. E. and Meyer, G. H. 1976. An up to date look at mould allergy. Ann. Allergy 37: 18-25.

Rajan, B.S.V., Nigam, S.A. and Shukla, R. K. 1952. A study of the atmospheric fugal flora at Kanpur. Proc. Indian Acad. Sci. B 35: 33-37.

Rajiv Kumar. 1982. Studies on the aerospora of Dehradun city. Abstr. Second International Conference on Aerobiology, 4-6 August, 1982, Seatle, U.S.A.

Rajkumar and Gupta, J. S. 1976. Seasonal and diurnal fluctuations in the air spora over a potato field. Indian Phytopathol. 29: 181-185.

Ramalingam, A. 1971. Air spora of Mysore. Proc. Indian. Acad. Sci. B 74: 227-240.

Raper, K. B. and Fennel, D. I.. 1973. The genus *Aspergillus*. Robert E. Kreiger, New York.

Raper, K. B. and Thomas, C. 1968. A Manual of Penicillia. Hafner Publishing Co., New York and London.

Rati, E. and Ramalingam, A. 1976. Airborne aspergilla at Mysore. Asp. Allergy Appl. Immunol. 9: 139-149.

Reyes, A.C., Vyenco, R.R., Punsalang, A.P., Jr. Fontanilla, E. and Lopez, S. 1970. Atmospheric mold spore count in Cebu City, Philippines. South East Asian J. Trap. Med. Pub. Hlth. 1: 263-269.

Richards, M. 1954. Atmospheric mould spores in and out of doors. J. Allergy 25: 429-439.

Richards, M. 1956. A census of mould spores in the air over Britain in 1952. Trans. Br. Mycol. Soc. 39: 431-441.

Robin, C. 1871. Traite du microscope. Paris.

Sandhu, D.K. and Singh, S. 1985. Airborne thermophilous fungi at Amritsar, India. Trans. Br. Mycol. Soc. 84: 41-45.

Sandhu, D.K. and Waraich, M.K. 1981. Airborne yeasts in Amritsar (India). Antonie Leewenhoek. J. Microbiol. 47: 571-576.

Sandhu, D.K., Shivpuri, D.N. and Sandhu, R.S. 1964. Studies on the airborne fungal spores in Delhi. Their role in respiratory allergy. Ann. Allergy 22: 374-384.

Sankara Rao, S. 1988. Studies on fungal air spora of crop fields. M. Phil Dissertation, Andhra University, Waltair.

Sarna, N. J. and Govil, C. M. 1979. Survey of air spora of Jaipur city. J. Indian Bot. Soc. 58: 215-225.

Shenoi, M. M. and Ramalingam, A. 1975. Circadian periodicities of some spore components of air at Mysore. Arogya J. Health Sci. 1: 154-156.

Shenoi, M. M. and Ramalingam, A. 1976a. Aerial dissemination of grain smut of sorghum (*Sphacelotheca sorghi*). Proc. Indian Natn. Sci. Acad. B 42: 194-204.

Shenoi, M. M. and Ramalingam, A. 1976b. Air spora of a sorghum field at Mysore. J. Palynol.

12: 43-54.

Shenoi, M. M. and Ramalingam, A. 1979. Epidemiology of sorghum downy mildew. II Circadian and seasonal periodicities in conidia and oospores. Proc. Indian Acad. Sci. B 88: 95-102.

Shivpuri, D.N. 1982. Studies in allergy to fungi in India. Asp. Allergy Appl. Immunol. 15: 19-30.

Singh, A. B. and Babu, C.R. 1983. Air borne fungal spora of Delhi. Indian J. Chest Dis. Allied Sci.1: 31-35.

Singh, N.I. and Baruah, H. K. 1979. Effect of air temperture, relative humidity and rainfall on the prevalence of air borne spores. Asp. Allergy Appl. Immunol. 12: 90-100.

Singh, N.I., Baruah, P. and Baruah, H. K. 1981. Seasonal periodicity of *Cladosporium* species in the air of Shillong. J. Palynol. 17: 111-120.

Singh, B.P., Mukerji, P.K. and Nath, P. 1979. Allergic significance of some fungi prevailing in the respiratory tract of allergy patients- a preliminary study. J. Asthma Res. 16: 51-53.

Singh, B.P., Nath, P. and Mukerji, P. K. 1980. Air borne fungi of Lucknow city- a study of two sites. Indian. J. Chest Dis. Allied. Sci. 22: 199-203.

Solomon, W. R. and William, J.A. 1970. A simplified application of the Andersen sampler to the study of air borne fungus particles. J. Allergy 45: 1-13.

Southworth, D. 1974. Introduction to the biology of air borne fungal spores. Ann. Allergy 32: 1-23.

Sreeramulu, T. 1967. Aerobiology in India. J. Scient. Industr. Res. 26: 474-480.

Sreeramulu, T. 1969. Final technical report of the U.S.D.A., PL. 480 Research Project "Studies on the incidence in the air over sugarcane fields of fungus spores pathogenic to the crop". pp.1-57.

Sreeramulu, T. 1970. Conidial dispersal in two species of *Cercospora* causing tikka leaf spots on groundnut (*Arachis hypogea*). Indian J. Agric. Sci. 40: 173-178.

Sreeramulu, T. 1971. Aeromycological observations and their implications in the epidemiology of some diseases of sugar cane. Proc. Indian Nat. Sci. Acad. B 37: 506-510.

Sreeramulu, T. and Ramalingam, A. 1962. Notes on airborne *Tetraploa* spores. Curr. Sci. 31:121-122.

Sreeramulu, T. and Ramalingam, A. 1963. Spore content of air over paddy fields. II. Changes in a field near Visakhapatnam from November 3, 1959 to January 9, 1960. Proc. Nat. Acad. Sci. B 33: 423-428.

Sreeramulu, T. and Ramalingam, A. 1964. Some short period changes in the atmospheric spore content associated with changes in the weather and other conditions. Proc. Indian Acad. Sci. B 59: 154-172.

Sreeramulu, T. and Ramalingam, A. 1966. A two year study of the air spora of paddy field near Visakhapatnam. Indian J. Agric. Sci. 36: 111-132.

Sreeramulu, T. and Seshavataram, V. 1962. Spore content of air over paddy fields. I. Changes in a field near Pentapadu from 21st September to 31st December, 1957. Indian Phytopathol.

15: 61-74.

Sreeramulu, T. and Vittal, B.P.R. 1966a. Periodicity in the air borne spore of the rice false smut fungus *Ustilaginoidea virens*. Trans. Br. Mycol. Soc. 49: 443-449.

Sreeramulu, T. and Vittal, B. P.R. 1966b. Some aerobiological observations on the rice stackburn fungus, *Trichoconis padwickii*. Indian Phytopathol. 19: 215-221.

Sreeramulu, T. and Vittal, B.P.R. 1969. Concentration of *Acellinopsis sacchar*i in the two heights. Indian Phytopathol. 22: 301-305.

Sreeramulu, T. and Vittal, B.P.R. 1970. Incidence and spread of red rot lesions on midribs of sugar cane leaves. Pl. Dis. Reptr. 54: 226-231.

Sreeramulu, T. and Vittal, B.P.R. 1971. Periodicity in the Uredospore content of air within and above the sugar cane field. J. Indian Bot. Soc. 50: 39-44.

Sreeramulu, T. and Vittal, B.P.R. 1972. Spore dispersal of sugar cane smut (*Ustilago scitaminea*). Trans. Br. Mycol. Soc. 58: 301-312.

Sreeramulu, T., Vittal, B.P.R. and Ramakrishna, V. 1971. Aerobiology of *Cercospora koepkei kriiger* causing the yellow spot disease of sugar cane. Indian J. Agric. Sci. 41: 655-662.

Stakman, E. C. and Christensen, C. M. 1946. Aerobiology in relation to plant disease. Bot. Rev. 12: 205-253.

Stix, E. 1977. Pilzsporengehalt de luft inMunchem. Munch Med. Wschr. 119: 78-82.

Storm Van Leeuwen, W. 1924. Bronchial asthma in relation to climate. Proc. R. Soc. Med. 17: 19-26.

Subba Reddi, C. 1970. A comparative survey of atmospheric pollen and fungus spores at two places twenty miles apart. Acta Allergol. 25: 189-215.

Subba Reddi, C. 1973. A preliminary study of sporangial population of Sclerospora graminicola in the air over a bajra field. Indian Phytopathol. 26: 157-159.

Subba Reddi, C. 1974. Circadian periodicity in the incidence of air borne spores of *Sporomia, Pithomyces* and *Spegazzinia*. Curr. Sci. 43: 85-86.

Subba Reddi, C. and Atluri, J. B. 1977. Circadian periodicity in air borne sores of *Dendryphiella, Pestalotia* and *Periconia*. Indian Phytopathol. 30: 570-572.

Subba Reddi, C., Atluri, J. B., Reddi, E.U.B. and Reddy, J. 1979. Circadian periodicity in the atmospheric content of *Drechslera, Periconia* and *Phyllachora* over a lawn. Kavaka 7: 27-31.

Subba Reddi, C., Atluri, J. B. and Reddi, E.U.B. 1980. Periodicity in aeciospores release in *Puccinia blepharidis*. Indian Phytopathol. 33: 495-498.

Subramanian, C.V. 1971. Hyphomycetes. I.C.A.R. New Delhi.

Targow, A.M. 1966a. Studies on atopic skin-test reaction: I. Survey of reactions to the fungi evidence for existence of two different group allergies. Ann. Allergy 24: 120-127.

Targow, A.M. 1966b. Studies on atopic skin-test reactions: II. Fallacies that underlie the concept that fungi contain species-specific allergens. Ann. Allergy.

Taylor, R. L. and Mcfadden, A.W. 1962. Survey of air borne mold flora in Panama. Mycopath. Mycol. Appl. 17: 159-164.

Thompson, D. 1854. Appendix to the report of the committee scientific enquiry in relation to cholera epidemic of 1854.

Tilak, S. T. and Babu, M. 1981. Aerobiological approach to leaf spot disease bajra (*Pennisetum typhoides*). Indian J. Bot. 4: 87-90.

Tilak, S.T. and Babu, M. 1982. Incidence of *Cunninghamella* spores in the atmosphere. Pollen Res.1: 55-58.

Tilak, S. T. and Bhalke, S. P. 1978. Aerobmycology at Aurangabad II Deuteromycetes. Indian J. Bot. 1: 113-115.

Tilak, S.T. and Bhalke, S. P. 1979. Aeromycology at Aurangabad III Hyphal fragments. Advancing Frontiers of Mycol. Plant Pathol. 51-54.

Tilak, S.T. and Kulkarni, R.L. 1975. Atmospheric concentration of *Rhizopus* spores. Biovigyanam 1: 91-94.

Tilak, S.T. and Kulkarni, R.L. 1978. Rust and smut spore content of air above a sugar cane field. Biovigyanam. 4: 103-108.

Tilak, S.T. and Kulkarni, R.L. 1980. Addition to the fungal flora of the air. Indian Phytopathol. 34: 69-71.

Tilak, S.T. and Mane, D. A. 1981. Aerobiological approach to rust disease of Bajra (*Pennisetum typhoides*). Proc. Nat. Conf. Env. Biol. 1981.

Tilak, S. T. and Srinivasulu, B.V. 1971. Air spora of Aurangabad II Ascospores. Indian Phytopathol. 24: 740-743.

Tilak, S. T., Talib, S. H. and Babu, M. 1980. Allergenicity of certain air borne fungal spores of hospital ward. Environment India 3: 27-30.

Tilak, S. T., Saibaba, M., Pillai, S.G. and Bhasale, S.S. 1981. Aeromycology of Aurangabad V Basidiomycetes. Proc. Nat. Cont. Env. Bio. 49-54.

Tkachyk, S. J. and Khan, R. S. 1977. Airborne mold survey of Edmonton. J. Asthma Res. 14: 103-106.

Turner, P. D. 1966. The fungal air spora of Hong Kong as determined by the agar plate method. Trans. Br. Mycol. Soc. 49: 255-267.

Tyndall, J. 1881. Essays on the floating matter of the air in relation to putrefaction an infection. Longmans, London.

Varma, K.V. 1986. Air spora of Bhimavaram. Ph.D. Thesis, Andhra University, Waltair.

Vijay Kumar, A., Ramgopal, G. and Rajeswara Rao, G. 1984. Aerobiological studies at Tirupathi. Abst. SV-21, Fourth Indian Palynological Conference, March 19-21. 1984, Andhra University, Waltair.

Vishnu-Mittre and Khandelwal, A. 1973. Air borne pollen grains and fungal spores at Lucknow during 1969-70. Palaeobotanist 22: 177-185.

Vittal, B. P.R. and Krishnamoorthi, K. 1981. Air spora of an agricultural farm in Madras, India. Grana 20: 61-64.

Vittal, B.P.R. and Krishnamoorthi, K. 1984. A one year volumetric survey of air mycoflora of Madras city. Abst. Fourth Indian Palynological Conference, March 19-21, 1984. Andhra University, Waltair.

Vittal, B.P.R. and Krishnamoorthi, K. 1988. A census of air borne mold spores in atmosphere in the city of Madras. India. Ann. Allergy 60: 99-101.

Vittal, B.P.R. and Krishnamoorthi, K. 1989. Circadian and seasonal periodicities of some mold allergens in the atmosphere of Madras city. India. Proc. Indian Acad. Sci. (Plant Sci) 99: 247-251.

Vittal, B.P.R. and Ponnuswamy, P. 1979. A preliminary study of the atmospheric fungal flora of Madras. 7: 79-82.

Vittal, B.P. R. and Ponnuswamy, P. 1982. J. Microbiol. 22: 79-81.

Vittal Murthy, K. P.R. 1981-84. Urban air quality assessment for Visakhapatnam and its impact on environment. Technical Report submitted to the Department of Environment, Government of India.

Vlodavets, V. V. 1956. Mould fungi in the atmospheric air of Moscow. Nature (Moscow) 45: 95-97.

Wadhwani, K. 1979. Air borne fungi flora of Lucknow (University area). J. Palynol. 15: 105-110.

Wadie, S. S. and Deshpande, K .S. 1980. Aeromycology of cotton fields. Sampling through cylinder spore trap. Proc. Indian Acad. Sci. (Pl. Sci.) 89: 15-22.

Warcup, J. H. 1959. Distribution and detection of root disease fungi. In: (Eds.) C.S. Holton, G.W. Fischer, R.W. Fulton, Helen Hart and S. E. A. McCallan, pp. 317-226, Plant pathology. The University of Wisconsin Press, Madison, Wisconsin.

Ward, H.M. 1882. Researchers on the life history of *Hemileia vastatrix*, the fungus of coffee leaf disease. J. Linn. Soc. (Bot.) 19: 299-335.

Werff, Vander, P. J. 1958. Mould fungi and bronchial asthma. Vol. I. Eds. H.E. Stenfert Kroese and N.V. Leiden.

Advances in Pollen Spore Research Vol. XXVIII (2010) : 131-141
Editor : Dr. A. J. Solomon Raju
Today & Tomorrow's Printers and Publishers, New Delhi - 110 002 (India)

POLLINATION BIOLOGY IN *IMPATIENS GARDNERIANA* WIGHT (BALSAMINACEAE)

Shrishail K. Kulloli, A. K. Sreekala,

A. G. Pandurangan and R. Ramasubbu

Plant Systematics and Evolutionary Science Division

Tropical Botanic Garden and Research Institute

Palode, Thiruvananthapuram 695 562

ABSTRACT

A study on the pollination biology including breeding system and pollinator behaviors in *Impatiens gardneriana* Wight, an endemic and horticulturally promising balsam from Western Ghats has been conducted. It flowers during July to December with optimum flowering in the month of September. Anthesis has occurred during the night time between 0100 and 0300h and anther dehisced one day before anthesis. Pollen grains are round, light yellow in color and 33.16 µm in diameter. Flowers offer pollen and nectar to the visitors which include honeybees (45%), butterflies (30%), hawk moths (15%) and flies (10%). Nectar is 2.5-4 ± 0.52 µl per flower; production begins soon after anthesis and continues up to 3 h. Pollen fertility is 78% with 82% viability and 69% of germination capacity. Stigma receptivity is up to 62% and pollen tube grows up to 943 µm long. Seed germination was limited to 25%. Xenogamy increases fruit set to 64%. Therefore, it is suggested that to external agents are required to increase its population under natural conditions.

Keywords: *Impatiens gardneriana*, pollination, phenology, protandry, Balsaminaceae, Western Ghats.

INTRODUCTION

Impatiens is the largest genus in the family Balsaminaceae (Sreekala et al. 2007; 2008). Its members are popularly known as balsams, jewel weed, snap weed or touch-me-not (Willis 1973) busy-lizzy (Caris et al. 2006). There are about 1,000 species distributed worldwide (Grey-Wilson 1980, Clifton 2000, Yuan et al. 2004, Caris et al. 2006), most of which occur in montane in subtropical regions from India to South-East Asia (Tian et al. 2004). It is essentially a sub-cosmopolitan and phytogeographically a unique genus which has its greatest development in the Indian region (Bhaskar 1981). In India, the concentration of *Impatiens* species is remarkably local and occurs in three well defined regions viz., the Western Himalayas, hills of north eastern states (Sikkim, Arunachal Pradesh and Darjeeling district in West Bengal) and the Western Ghats (Hills south of Tapti river in the north to Kanyakumari in the south, along the west coast of India). Two hundred and six species have been found in India (Vivekananthan et al. 1997, Vishwanathan and Manikandan 2003, Bhaskar 2006, Sreekala et al. 2008a) which represents approximately 20% of *Impatiens* species globally. Most of them are found in the Western Ghats and Western Himalayas and are endemic to the country or are restricted to a number of provinces (Kulloli et al. 2009, 2009a). The geographical distribution of *Impatiens* is very localized and endemic (Tian et al. 2004). In Peninsular India there are 92 species of *Impatiens* currently reported, out of which 86 are endemic and confined to the Western Ghats (Nair 1991, Rajalal and Pandurangan 1997, Kulloli et al. 2009). Due to their restricted distribution and island biogeography, nearly 30 species are already in threatened category with uncertain future (Vajravelu and Daniel 1983, Pandurangan and Pushpangadan 1997, Sreekala et al. 2007, Kulloli et al. 2009a).

Though the ideal climatic conditions prevailing in the Western Ghats region provide suitable habitat for the *Impatiens*, their populations are rapidly declining due to habitat destruction and degradation resulting fragmentation of populations in their narrow environmental niche by anthropogenic pressures where reproductive constraints increases by elimination or reduction of pollinators. The pollination biology is a prerequisite in plant breeding for obtaining better yields of such endemic plants. It also offers vital information to the plant hybridization programme. In this context, a detailed investigation on the pollination biology of *Impatiens gardneriana* Wight an endemic balsam of the Southern Western Ghats has been undertaken to find out causes of rarity reasons for dwindling populations in the natural habitats, also to throw light on different methods of pollination. This will facilitate to increase population of such species by effective conservation measures by the planners, foresters, conservationists and naturalists.

MATERIALS AND METHODS

Impatiens gardneriana was collected from Periya, Thirunelli, Banasursagar of Wayanad district and Silent Valley National Park of Palakkad district, Kerala, India. The plants were grown in the field garden of Tropical Botanic Garden and Research Institute, Palode, Thiruvananthapuram for further comprehensive investigation. Floral morphology of the flower was studied in detail with field dissection microscope. An extensive field exploration has been conducted in the Wayanad district to find out the flowering period and floral nature of the candidate species. Flower colour, odour and nectar were observed visually. Anthesis and anther dehiscence were observed using a fluorescent lamp at night and a hand lens following the method of Reddi and Janaki Bai (1981), Mathur and Mohan Ram (1986). The pollen production per flower was estimated as per the method suggested by Dafni et al. (2005) and pollen-ovule ratio (P/O) was calculated following Cruden's (1977) method.

Pollen fertility and viability was assessed by acetocarmine glycerin staining techniques and FCR (fluorochromatic reaction) test using FDA (fluorocein diacetate) respectively suggested by Shivanna and Rangaswamy (1992). To have knowledge about *in vitro* pollen germination, freshly collected and uncontaminated pollen from immediately dehisced anther was put onto to the sucrose solution of different concentrations (5, 10, 15, and 20%) and Brewbaker's medium. The prepared slides were kept within Petri dishes lined with moist filter paper for incubation. After the stipulated period, the percentage of pollen germination and pollen tube elongation was noticed under light microscope (Lieca DME, Germany) and calculations were made following the method of Shivanna and Rangaswamy (1992). Stigma morphology and receptivity was studied with the help of hand lens and, pollination experiments and with the help of Scanning Electronic Microscopy (SEM- SU-70 Hitachi). *In vivo* pollen germination studies were conducted as per the method suggested by Joshirao and Saoji (1989) and with the help of Scanning Electronic Microscopy (SEM- SU-70 Hitachi). To measure the amount of nectar per flower at different time intervals (1 h), a total of 50 flowers were collected for this experiment. Measured the length of nectar column of capillary tube (for volume calculation) and also measured the concentrations with hand held refractometer (0-35%, Erma, Japan).

During all the visits made to the plants under observation, representative samples of flower visitors were collected using insect net and later identified by our research group and standard insect taxonomy manual (Borror and White 1970, Carr 1975, Bland and Jacques 1978, Senagupta and Tapan 2005). Further, the type of food collected by different pollinators and foraging period was recorded by close observation. Tests for autogamy, geitonogamy and xenogamy were conducted on 50 flowers distributed in 10 plants of five populations for each mode of reproduction and fruit sets recorded.

RESULT AND DISCUSSION

Phenology: The plant sets full vegetative growth following monsoonal rains in June and starts flowering during late July (Fig. 1A and Table 1). Flowering is synchronous in the habitat. *Impatiens tomentosa* Heyne ex Wight & Arn., *I. viscosa* Bedd. and *I. chinensis* Barnes are the other *Impatiens* species that grow and flower along with *Impatiens gardneriana*. It is an annual succulent herb reaching a height of 40cm-1m. Flowers are born in axillary fascicles, sometimes binate in nature with attractive bright pink petals (Fig. 1B and Table 1). The lip is cymbiform ending with slender and curved spur (Fig. 1D). Capsules are ellipsoidal and glabrous; produce very few hairy seeds (Fig. 1F). In *I. gardneriana*, flower buds initiated in the second week of July and extended up to December (Table 1) with maximum bloom in October. The plants were in flower for a total of 180 days in a year. The flower buds take 4-6 days from initiation to full bloom (Fig. 1C). The average life span of each flower is 2-3 days. Flower opening commenced between 0100-0300h. Anther dehiscence was observed one day before flower opening, which in turn confirmed the protandrous condition of the flower (Table 1). Sreekala et al. (2007, 2008) also reported parallel findings in *Impatiens hensloviana* Arn. and *Impatiens coelotropis* Fischer. Bhaskar and Razi (1974) had reported that majority of wild balsams grown in the high altitude areas are night blooming and have wide range of timing with regard to anthesis and pollen germination. Study of anthesis and anther dehiscence is vital for subsequent dispersal of pollen grains into the atmosphere (Bhattacharya and Mandal 2000). The mean number of pollen grains and ovules per flower are ±20,300 and 18 respectively. Hence, the pollen-ovule ratio had been worked out as 1990 pollens per ovule (1990:1). This higher pollen-ovule ratio suggests out-breeding nature of the species. Pollen productivity depends on anther length, pollen grain size and mode of anther dehiscence (Linskens 1992, Reddi and Reddi 1986) and total pollen output is related to fruit and seed setting.

Table 1. Floral characters of *Impatiens gardneriana*

Floral Characters	Observations
Inflorescence	Terminal raceme
Flower type	Hermaphrodite, zygomorphic
Flower colour	Bright Pink
Flowering period	July-December
Odour	Mild

Nectar	Present (2.5-4µl ±0.52)
Flower opening time	0100-0300h
No. of anthers/flower	5
Anther dehiscence time	0230-0330h (One day before
Anther dehiscence mode	anthesis)
Average no. of pollens/anther	Apical slit
Mean no. of pollen	4460±30
grains/flower	20,300±50
Mean no. of ovules/flower	18
Pollen-ovule ratio	1990:1
Pollen type	3-colporate
Pollen size	±33.16µm
Pollen shape	Round
Stigma type	Wet and non-papillate
Stigma receptivity %	62%

Pollen fertility, viability and in vitro pollen germination:

In *Impatiens gardneriana,* the pollen grains are round and on an average of ±33.16ìm in diameter (Table 1). The acetocarmine-glycerin staining technique revealed that, 78% of pollen grains were only fertile. Pollen viability by FCR test also confirmed that 82% pollen grains were viable on the day of anthesis. *In vitro* pollen germination by using different concentration of sucrose revealed that, 69% pollen grains were germinated and achieved tube length of 724µm in 10% sucrose medium. Germination percentage was significantly low in higher concentrations of sucrose medium. Highest percentage of pollen germination of 74% along with 801µm tube length was noticed in Brewbaker's medium (1963) after 4h of incubation. It contains sucrose which is the best carbohydrate source for pollen germination having its function in maintaining osmotic pressure of the medium and acting as a substrate for pollen metabolism (Johri and Vasil 1961; Shivanna and Johri 1985). Pollen viability is maximum on the day of anthesis and gradually decreased on successive days after anthesis. This observation is analogous to that of *Impatiens reptans* Moon in China (Tian et al. 2004) and *I. coelotropis* Fischer in India (Sreekala et al. 2008).

Stigma receptivity and in vivo pollen germination:

Receptivity of stigma is a critical factor for successful completion of post pollination events in the flowering plants. The stigma is wet and non-papillate type and receptive up to 62% on the first day of flower opening with 62% *in vivo* germinating pollens along with 943µm long pollen tube on the stigmatic surface. Subsequently, the receptivity and *in vivo* germination percentages were decreased on successive days after anthesis. Pollen fertility and pollen tube growth on the stigma surface and the style are very sensitive to temperature charges. After being successfully placed on stigma surface the pollen undergoes hydration. Subsequent to hydration, pollen wall proteins are release on the stigmatic surface (Heslop-Harisson et al. 1975) and the stigmatic pellicle acts as a receptor of pollen wall proteins (Joshirao and Saoji 1989). Thus the pollen grains are recognized and begin to germinate on the stigmatic surface.

Nectar analysis:

In *I. gardneriana,* the nectar secretion begins an h before anthesis and continues even after anthesis for ca. 3h. The fully open flower holds 2.5-4µl ±0.52 of nectar. A sugar concentration of nectar at anthesis is 15% and later it varies between 16 and 20%, probably due to water evaporation during bright sunlight h. Nectar in *Impatiens* species are relatively minor but long spurred species shows fairly measurable amount of nectar e.g. the length of the spur ranges from 2-5cm. The amount of nectar present in each flower is extremely vital for pollinators that determine the rate of visit by each potential pollinator to the individual flower.

Pollinators and their behavior:

Flower opening commenced between 0100-0300 h and anther dehiscence was observed one day before flower opening. The major visitors of *Impatiens gardneriana* are honeybees, butterflies, hawk moths and flies (Table 2). The ants are also found on vegetative and floral parts, moving from one flower to another but not effected pollination and foraged for nectar only. The foraging period of honeybees (*Apis cerana* Fabricius, *Apis dorsata* Fabricius and *Apis indica* Fabricius) is mainly in day time around 0630-1600h. They carried the pollen loads on the head, thorax and hind legs and transferred the pollen grains from one flower to another of the same plant or flowers of different plants and thus promoting geitonogamy and xenogamy. They actively forage for both pollen and nectar. Pollen bee (*Trigona iridipennis* Smith) is the most abundant visitor and forage only pollen during its visit and simultaneously effect the pollination (Fig. 1D). During the night time (2230-0300h) a species of hawk moth (*Macroglossum*

Table 2. Pollinators and their foraging behaviours in *Impatiens gardneriana*

Order	Family	Genus	Species	Foraging nature	Foraging time	Foraging Hour (h)
Hymenoptera	Apidae	*Apis*	*cerana*	N and P*	Day	0730-1130h
		Apis	*indica*	N and P	Day	0630-1600h
		Apis	*dorsata*	P	Day	0700-1200h
		Trigona	*iridipennis*	N	Day	0700-1400h
Lepidoptera	Papiliodae	*Pachliopta*	*aristolochiae*	N	Day	0730-1500h
	Danaidae	*Euploea*	*core*	N	Day	0800-1430h
		Danaus	*chrysippus*	N	Day	0630-1000h
	Sphingidae	*Macroglossum*	*variegatum*	N	Night	2230-0330h

*N- Nectar, *P- Pollen

variegatum Rothschild and Jordan) was actively visited the flowers and spending an average of 2 sec in each flower. Butterflies (*Pachliopta aristolochiae* Fabricius, *Euploea core* Cramer and *Danaus chrysippus* Linn.) are active during day time (0630-1400h) and they visited flower for nectar and spending an average of 3-5s per flower (Table 2). Of the total visits, bee made 45%, butterflies 30%, moths 15% and flies 10%. All the insects are pollen carriers and their frequent inter plant movement facilitate cross pollination. However, the visits of the insects are not sufficient to pollinate all the flowers in the populations. It was calculated that 52 insects are visiting nearly 1000 flowers per h. This may lead to the poor fruit set in the natural condition.

Honeybees are the major pollinators of *Impatiens gardneriana*. This observation is corroborated with the findings of Sreekala et al. (2008) in *Impatiens coelotropis*. It is well known that the flowers of *Impatiens* have enormous diversity and different pollinators. In tropical regions of Indian subcontinent the *Impatiens* are pollinated by honeybees, butterflies and moths (Sreekala et al. 2007; 2008), whereas in subtropical regions of Africa the *Impatiens* are pollinated by humming birds as well as by insects (Grey-Wilson 1980). In temperate zones, pollinators are bumble-bees and humming birds only (Rust 1977; 1979; Heinrich 1979; Kato et al. 1989). Hence, in different climatic regions, species of pollinators vary with respect to geographic set-up.

Table 3. Fruit set in different modes of pollination in *Impatiens gardneriana*

Sl.	Treatments	No. of flowers pollinated fruits	No. of flowersset	% of fruit set
1	Autogamous self pollination (B)	50	0	0
2	Geitonogamy (S)	50	12	24
3	(O) xenogamy	50	32	64
4	Natural pollination (N)	50	24	48

Breeding system:

In natural pollination, 38% fruit set was observed. Fruit set was not observed in autogamous self-pollination. However 24% fruit set was observed in geitonogamy and 64% in xenogamy respectively (Table 3). The seed germination percentage in both natural and lab condition was only 25%. Capsules developed through xenogamy had more seeds than capsules produced by natural pollination, this has been verified by artificial xenogamous pollination experiment which resulted 64% fruit set with more seeds, confirming that pollination is a limiting factor for fruit set in the *Impatiens gardneriana* (Fig. 1E and F).

CONCLUSION

We conclude that the species is poorly distributed in the Western Ghats mainly due to poor fruit set, low percentage of seed germination etc. Different pollination experiments demonstrated that manual cross-pollination especially xenogamy enhanced the rate of fruit set in *Impatiens gardneriana*, which strongly indicates that some external agents are required for effective transfer of pollen grains to ensure the successful pollination. In *Impatiens gardneriana* the extent of occurrence is restricted to less than 65 km^2 and hence, the availability of disjunct population is also a limiting factor for gene flow which greatly affects the sexual reproduction. All these causal factors continue to operate through out the year and these factors either alone or in combination with others are responsible for limited distribution in the natural habitats. However, such reproductive biological studies help to raise viable seedlings from improved seed output and thus a better recruitment in their natural conditions is required through xenogamic pollination aided by external agents.

ACKNOWLEDGEMENTS

The authors are thankful to Department of Science Technology, New Delhi for financial support. Thanks are also due to the Director TBGRI for inspiration and logistic support during the research programme.

REFERENCES

Bhaskar, V. and Razi, B. A. 1974. Nocturnal pollen germination in *Impatiens* (Balsaminaceae). *Curr. Sci.* 43: 626-628.

Bhaskar, V. 1981. The genus *Impatiens* L. in South India: endemism and affinities. – *Ind. For.* 107: 368-376.

Bhaskar, V. 2006. *Impatiens clavata* Bhasker sp. nov. - a new scapigerous balsam (Balsaminaceae) from Bisle Ghat, Western Ghats, South India. *Curr. Sci.*– 91: 1138-1140.

Bhattacharya, A. and Mandal, S. 2000. Pollination biology in *Bombax ceiba* Linn. *Curr. Sci.* 79: 1706-1711.

Bland, R. G. and Jacques, H. E. 1978. How to know the insects. Wm. C. Brown Camp Publishers, Iowa.

Borror, D.J. and White, R.E. 1970. A field guide to the insects. Houghton Mifflin Co., Boston, Mass.

Brewbaker, J. L. and Kwack, B. H. 1963. The essential role of calcium ion in pollen germination and pollen tube growth, *Am. J. Bot.* 50: 859-865.

Caris, P. L., Geuten, K. P., Janssens S. B. & Smets, E. F. 2006. Floral development in three species of *Impatiens* (Balsaminaceae). *Am. J. Bot.* 93: 1-14.

Carr, A. 1975. Rodale's color handbook of garden insects. Rodale Press, Emmaus, Penn.

Clifton, R. 2000. Geraniales species checklist series Vol. 6, Geraniaceae, Bentham and Hooker tribe VII. Balsamineae. – Dover: The Geraniaceae Group.

Cruden, R.W. 1977. Pollen-ovule ratios: a conservative indicator of breeding systems in flowering plants. *Evolution* 31: 32-46.

Dafni, A., Kevan, P. G. and Husband, B. C. 2005. Practical pollination biology. – Enviroquest Ltd. Cambridge, Ontario, Canada.

Grey-Wilson, C. 1980. *Impatiens* of Africa, A. A. Balkema, Rotterdam, Netherlands.

Heinrich, B. 1979. Resource heterogeneity and patterns of movements in foraging bumblebees. *Oecologia* 40: 235-245.

Heslop-Harrison, J., Heslop-Harrison, Y. and Barber, J. 1975. The stigma surface in incompatibility response. *Proc. Royal Soc., London.* Series B. 188: 287-297.

Johri, B. M. and Vasil, I. K. 1961. Physiology of pollen. *Bot. Rev.* 27: 325-381.

Joshirao, J. A. and Saoji, A. A. 1989. Studies on *in vivo* germination of pollen of some alkaloid bearing plants. *J. Palynol.* 25: 45-50.

Kato, M., Itino, I., Hotta, M., Abbas, I. and Okada, H. 1989. Flower visitors of 32 plants species in West Sumatra. Occ. Papers Kagoshima Univ. Res. Center South Pac. – 16: 15-31.

Kulloli, S. K., Ramasubbu, R. Sreekala, A. K and Pandurangan, A. G. 2009. Reproductive ecology of *Impatiens campanulata* Wight :A rare and endemic balsam of Southern Western Ghats. *Ecol. Env. Cons* 15(2): 33-37.

Kulloli, S. K., Sreekala, A. K. and Panduranagan, A. G. 2009a. Floral biology of *Impatiens trichocarpa* Hook. f., (Balsaminaceae) an endemic Balsam of Western Ghats. *Ind. J. Sci. Tech.* 2: 30-34.

Linskens, H. F. 1992. Sexual plant reproduction. In: Cresti, M. and Tezzi, A. (eds.). Springer-Verlog, Berlin. pp. 203-218.

Mathur, G. and Mohan Ram, H. Y. 1986. Floral biology and pollination of *Lantana camara. Phytomorphol*ogy 36: 79-100.

Nair, N. C. 1991. Endemism on the Western Ghats with special reference to *Impatiens* L. Proceedings of the symposium on rare, endangered and endemic plants of Western Ghats, –Kerala Forest Dept. (Wildlife Wing) Thiruvananthapuram Spl. Publ. No III, 93-102.

Pandurangan, A. G. and Pushpangadan, P. 1997. Causes of rarity and extinction of species in plants with reference to Western Ghats. – Frontiers in Plant Science. Om Sai Graphics, Hyderabad. Pp. 131-138.

Rajlal, B., Pandurangan, A. G. and Pushpangadan, P. 1996. Systematic studies on Balsaminaceae in Peninsular India. *J. Swamy Bot. Cl.* 13: 59-62.

Reddi, C. S. and Reddi, N. S. 1986. Pollen production in some anemophilous Angiosperms. *Grana* 25: 55-61.

Redii, C. S. and Janaki Bai, A. J. 1981. Floral biology of *Mimusops elengi*. *J. Bombay Nat. Hist. Soc.* 77: 471-475.

Rust, R. W. 1977. Pollination in *Impatiens capensis* and *Impatiens pallida* (Balsaminaceae). *Bull. Torrey Bot. Club.* 104: 361-367.

Rust, R. W. 1979. Pollination in *Impatiens capensis*: pollinators and nectar robbers. *J. Kansas Entom. Soc.* 52: 297-308.

Senugupta, T. M. and Tapan, A. 2005. *Insects of India.* Eastern book corporation, New Delhi. Pp. 284.

Shivanna, K. R. and Johri, B. M. 1989. The Angiosperm pollen-structure and function. –Wiley Eastern Ltd. New Delhi.

Shivanna, K. R. and Rangaswamy, N. S. 1992. Pollen biology- a laboratory manual. –Narosa Publ. House, New Delhi. pp. 35-36.

Sreekala, A. K., Pandurangan, A. G., Ramasubbu, R. and Kulloli, S. K. 2008. Pollination biology of *Impatiens campanulata* Wight. *Adv. Pollen Spore Res.* – 26: 9-19.

Sreekala, A. K., Pandurangan, A. G., Ramasubbu, R. and Kulloli, S. K. 2008a. Reproductive biology of *Impatiens coelotropis* Fischer a critically endangered balsam from the southern Western Ghats. *Curr. Sci.* 95: 386-388.

Sreekala, A. K., Ramasubbu, R., Kulloli, S. K. and Pandurangan, A. G. 2007. Pollination biology of *Impatiens hensloviana* Arn. (Balsaminaceae). *Ind. J. Bot. Res.* 3: 165-171.

Tian, J., Liu, K. and Hu, G. 2004. Pollination ecology and pollination system of *Impatiens reptans* (Balsaminaceae) endemic to China. *Ann. Bot.* 93: 167-175.

Vajravelu, E. and Daniel, P. 1983. Materials for a catalogue of threatened plants of India – Botanical Survey of India, Calcutta.

Vishwanathan, M. B. and Manikandan, U. 2003. A new species of Balsaminaceae, *Impatiens tirunelvelica*, from Peninsular India. *Bull. Bot. Surv. India* 45: 189-194.

Vivekananthan, K., Rathakrishnan, N. C., Swaminathan, M. S. and Ghara, L. K. 1997. Balsaminaceae. In: Hajra, P. K., Nair, V. J. and Daniel, P. (eds.), Flora of India (Maphighiaceae-Dichapetalaceae. Botanical Survey of India. Calcutta. 4: 95-229.

Willis, J. S. 1973. A Dictionary of Flowering Plants and Ferns. 8th revised edition by Airy Shaw, H. K. – Cambridge University Press, Cambridge. pp. 591-592.

Advances in Pollen Spore Research Vol. XXVIII (2010) : 143-148
Editor : Dr. A. J. Solomon Raju
Today & Tomorrow's Printers and Publishers, New Delhi - 110 002 (India)

PHENOLOGY AND REPRODUCTIVE BIOLOGY OF *PARTHENIUM HYSTEROPHORUS* L.

K. P. Singh, *Gunjan Dhakre and Vinay Daboriya

Department of Botany,

R.B.S. College, Agra 282002

**Department of Botany,*

School of life Sciences,

Dr.B.R.A .University, Agra

ABSTRACT

Phenology and reproductive biology of *Parthenium hysterophorus* was undertaken. There are two distinct generations, one in summer and the second in winter. The young plants are erect in summer while in winter they are spread in the ground. Numbers of inflorescence, flowers/inflorescence, pollen/anther, pollen fertility, *in vivo* pollen germination, fruits/plants, fruit-set and seed germination were recorded. The plants were found to be more reproductive in summer than in winter.

Key words: *Parthenium hysterophorus*, summer annual, winter annual.

INTRODUCTION

Parthenium hysterophorus L. (Asteraceae) is a native of Mexico and adjacent USA. In India, the weed was reported in 1955 (Rao, 1956) and is now spread to all the states of India (Dhawan *et al.* 1993). It invades all types of pasture lands and cause yield reduction in agricultural crops and in forage production in grasslands (Krishnamurthy *et al.* 1993; Saraswat 1993; Murthy *et al.* 1995). This antagonistic nature of *P. hysterophorus* is due to its higher reproductive capacity, effective utilization of resources and efficient seed dispersal (Saxena and Ramakrishnan 1982). In the present investigation an attempt has been made to study the phenology and reproductive biology of *P. hysterophorus.*

MATERIALS AND METHODS

Present investigation was undertaken on *P. hysterophorus* L. plants growing in different parts of Agra city. Floral morphology, number of pollen grains/anther, number of ovules /flower were recorded at regular intervals. Pollen fertility was checked by Alexander's (1980) staining technique, 1% tetrazolium chloride and FCR test (Shivanna and Rangaswamy, 1992). *In vivo* pollen germination studies were undertaken by method given by Shivanna and Rangaswamy (1992). Number of fruits/plant, number of seeds/fruit, fruit-set percentage and seed -set percentage was also observed.

RESULTS AND DISCUSSION

The results made on the phenology and reproductive biology of *Parthenium hysterophorus* L. are summarized in Table 1. The data in Table 1 clearly shows that *P. hysterophorus* grows as an annual in both winter and summer. The summer generation is initiated with monsoon rains in late June, while winter generation starts with low temperature in December. Plants in summer exhibit more luxuriant growth as compared to winter. Plants in summer are 100-150 cm tall while, plants in winter are about 50-100 cm in height. Young plants in summer are erect in form while, young plants of winter spread on the ground and cover more land area. In summer floral buds initiate in July on the plants attaining the height of 30-40 cm. However, young plants in winter with 5-10 cm height with a few leaves exhibit bud initiation in January-February. Flowering occurs in August in plants of summer. On the other hand, in winter, flowering takes place in March in very young plants. Such early allocation for reproduction when vegetative growth is just beginning has been called 'pessimistic strategy' by Cheplick and Quinn (1982).

Table 1: Phenology and Reproductive Biology of *Parthenium hysterophorus* L.

S.No.	Characters	Summer annual	Winter annual
1	Plant height	100-150 cm	50-100 cm
2	Time of Bud initiation	July	January-February
3	Time of flowering	August	March
4	Number of ray	5	5
5	floret/inflorescence	120-130	60-80
6	Number of disc	Unisexual pistilate	Unisexual pistilate
7	florets/inflorescence	Bisexual	Bisexual
8	Distribution of sex in ray floret	4	4
9	Distribution of sex in disc florets	Bifid	Bifid
10	Number of stamens /disc florets	Globular	Globular
11	Stigma: in ray floret	1	1
12	in disc floret	1	1
13	Number of ovule /ovary	175-220	150-160
14	Number of ovule/flower	800-850	600-650
15	Number of pollen/anther	80-95	70-80
16	Number of pollen /disc floret	80-90	60-70
17	Pollen fertility (%)	September	April
18	*In vivo* pollen germination	2500-3000	2000-2500
19	Time of fruit formation	90-95	80-90
20	Number of fruit /plant	1	1
21	Fruit set percentage	October-	May-June
22	Seeds/fruit	November	June-July
	Seed dispersal	December-January	90-99
	Seed germination	60-70	
	Seed germination percentage		

The inflorescence or head is pentangular in shape with two whorls of involucres. It measures 13.5 mm^2 in size in summer while, in winter, it is only 12.0 mm^2 in diameter. It is interesting to note that number of inflorescence/plant was high (200-250) in summer and low (150-200) in winter. Numerous, minute, white ray and disc florets are present in each head .The number of ray florets remain constant in each head during both seasons. However, the number of disc florets varies considerably. There are only 35-80 disc florets in mature inflorescence of winter annuals, while 120 -130 disc florets are observed in mature inflorescence of summer plants. Solbrig and Cantino (1975) reported that presence of a large number of flowers in a head is an adaptation that serves to attract insects to individually small flower or flowers which serve as functional equivalent to petal. In addition, shallow light and simple flowers provide a platform to be characteristic of fly pollinated flowers (Fagri and Pijl 1979).

In disc florets both reproductive organs i.e. androecium and gynoecium are present. Four stamens are recorded in each disc floret. In every anther about 175-220 pollen are recorded in summer annuals. Contrary to this, there are only 150-160 pollen per anther in winter annuals. There are 800-850 pollen anther in each disc floret of summer annual, while only 600-650 pollen were recorded per disc floret of winter annual. However, studies made by Meshram *et al.* (2001) explained that on an average, *P. hysterophorus* produces 812 pollen grains per flower. Total pollen production per anther was not dependent on the size of anther but on the habit of the plant (Meshram *et al.* 2001). Studies made by several workers suggest that more efficient the mechanism of pollen transfer, the lower will be the output of the plant (Wyatt 1984 and Cruden 2000). Therefore, in *Parthenium* pollen transfer does not seems to be very efficient and pollen is produced in large amount for its successful transfer mechanism. Lord (1980) considers pollen count/flower to be a function of the position the flower holds in the inflorescence, the environmental conditions under which the plant grows and the plants genetic make-up. Tilak and Patil (1983) also carried out extensive work on *Parthenium* pollen. They reported that maximum concentration of pollen (in air) was in the month of July ($655/m^3$) and minimum in December ($35/m^3$). According to Leushner (1974), high concentration of pollen was present when the temperature ranged between 27-37°C with high velocity, while concentration of pollen decreased with the increase in relative humidity. On the other hand, Tilak *et al.* (1981) studied that maximum concentration of *Parthenium* pollen was present in between 10.00-11.00 am and minimum between 04-06 am.

Ray florets in head inflorescence are unisexual with bicarpellary gynoecium and ovary is unilocular with a single basal anatropous ovule. On the other hand, in disc florets the ovary is rudimentary and fails to produce seed. Fruits are cypsella formed in September in the plants of summer. However, fruiting occurred in April. There are about 2500-3000 fruits per plant of summer while, 2000-2500 fruits develop per plant in winter. Rao (1992) also reported that about

5000 to 7000 cypsella per plant per year. Jayachandra (1971) reported that in India the weed continually produces flowers and fruits which emerge into a dense mat of seedlings after few showers. There is 90-95 % fruit set in plants of summer generation while only 80-90% fruit set is recorded in plants of winter annual. Seed fail to develop in disc florets while single seed is observed per fruit in ray floret. Seeds of *Parthenium* are white in colour, oval- angular in shape, smooth without any accessory organ. However, FcFadyen (1992) reported seeds with short wing like process and are without hooks or pappus. Seed dispersal occurs during the months of October-November in summers while in winters seed dispersal takes place in the months of May – June. Seeds are distributed by floating, or still or flood water, or in mud adhering to animals, vehicles and machinery (Auld *et al.* 1982, 1983). Seeds of *Parthenium* germinate in the month of December to give rise the plants of winter generation, however seeds germinate in June to form summer annuals. Seed germination percentage was 60-70 for winter generation. However, in June and July the seed germination was 90-99 %. Sukhada (1977) reported that seeds of *Parthenium hysterophorus* remain viable up to two years.

REFERENCES

Alexander, M.P. 1980. A versatile stain for pollen fungi, yeast and bacteria, Stain Technology 55:13-18.

Auld, B.A., Hosking, J. and Fayden, R.E.1982, 83. Analysis of the spread of tiger pear and *Parthenium* weed in Australia, Aust. Weeds 2:56-60.

Cheplick, G.P. and Quinn, J.A.1982. *Amphicarpum purshil* and the pessimistic strategy in amphicarpic annuals with subterranean fruit. Oecolgia 52:327-332.

Crden, R.W.2000. Pollen grains: why so many? Pl. Syst. & Evol. 222: 143-166.

Dhawan, S.R., K.R.Aneja and Poonam, Dhawan. 1993. *Parthenium hysterophorus* L. -The Danger weed and its control. Biome 6(2): 117-122.

Faegri, K. and Pijl, L. Vander1979. The principles of pollination ecology, Pergamon Press, Oxford.

Fayden, R.C., 1992. Biological control against *Parthenium* weeds in Australia. Crop. Protection 11:400-407.

Jayachandra, 1971. *Parthenium* weed in Mysore state and its control. Curr.Sci. 40:358.

Krishnamurthy, M.K.Munegowda and Jagdish, A., 1993. Ecological weed survey in Karnataka Proc.lnt.Sym.Integrated Weed Management for Sustainable Agriculture, Indian Soc. of Weed Science, HAU, Hisar, and 1:13-52.

Leushner, R.M. 1974 Pollen calendars for Switzerland. In: Atlas of European allergenic pollens (ed. Charpin et al.), pp.193-199.

Lord, E.M.1980.Intra-inflorescence variability in pollen /ovule ratios in the cleistogamous species, *Leminum amplexicaule* (Labiatae). Am. J. Bot. 67(4): 529-533.

Meshram, B.M., Meshram, S., Patil, V.B. and Mulay, J.R.2001.Pollination Biology and Incidence of Allergy due to *Parthenium*. Int. J. Aerobiol. 14(172):43-44.

Murthy, BG., Prathiba, N. C. and Thammaiah, N. 1995. Study on allelopathic effect of *Parthenium* on sunflower and sorghum. The World Weeds 2(3&4):161-164.

Rao, R.S., 1956. *Parthenium hysterophorus* Linn. A new record for India. J. Bombay. Nat. Hist. Soc. 45:218-220.

Rao, V.S. 1992. Principles of Weed Science (A Reference cum Text Book), Oxford & IBH Publishing Co. Pvt. LTd., New Dehli, p. 540.

Saraswat, V.N. 1993. Major weeds of Indian Agriculture-Their distribution and ecology. Proc. Int. Sym. Integrated Weed Management for sustainable Agri. Ind. Soc. Of Weed Science, HAU, Hisar 1: 35-41.

Saxena, K. G. and Rangaswamy, N. S. 1992. Pollen biology: A laboratory manual, Springer Verlag Heidelberg, Germany.

Shivanna, K. R. and Rangaswamy, N. S. 1992. Pollen biology: A laboratory Manual Springer–Verlag, Heidelberg, Germany.

Solbrig, O.T. and Cantino, P.D. 1975.Reproductive adaptations in *Prosopis* (Leguminosae, Mimosoideae). J. Arn. Arbo. 56 (2):185-210.

Sukhada, D.K. 1977. Agro-physiological impact of the Allelopathic Influence of *Parthenium hysterophorus* L. Ph. D. Thesis, Bangalore University.

Tilak, S.T., Tilak, S.H. and Bhansale, S.S. 1981. Inhalent allergens inside the hospital ward. Proc. Nat. Conf. Env. Biol. Aurangabad 113-116.

Tilak, S.T. and Patil, B.Y.1983. The status of *Parthenium* pollen as aeroallergen. J. Biol. Res 3:12-15.

Wyatt, R. 1984. Evolution of self-pollination in granite outcrop species of *Arenaria* (Caryophyllaceae). Reproductive effort and pollen-ovule ratios. Systematic Botany 9:432-440.